The Global Challenge of Innovation

The Global Challenge of Innovation

Basil Blackwell and Samuel Eilon

Butterworth-Heinemann Ltd
Halley Court, Jordan Hill, Oxford OX2 8EJ

 PART OF REED INTERNATIONAL BOOKS

OXFORD   LONDON   GUILDFORD   BOSTON
MUNICH   NEW DELHI   SINGAPORE   SYDNEY
TOKYO   TORONTO   WELLINGTON

First published 1991

© Basil Blackwell and Samuel Eilon 1991

All rights reserved. No part of this publication
may be reproduced in any material form (including
photocopying or storing in any medium by electronic
means and whether or not transiently or incidentally
to some other use of this publication) without the
written permission of the copyright holder except in
accordance with the provisions of the Copyright,
Designs and Patents Act 1988 or under the terms of a
licence issued by the Copyright Licensing Agency Ltd,
33–34 Alfred Place, London, England WC1E 7DP.
Applications for the copyright holder's written permission
to reproduce any part of this publication should be addressed
to the publishers.

British Library Cataloguing in Publication Data
Blackwell, Basil
  The global challenge of innovation.
  1. Economic conditions. Innovation
  I. Title II. Eilon, Samuel
  338.064

ISBN 0 7506 0077 2

Photoset by Redwood Press, Melksham, Wiltshire
Printed and bound in Great Britain.

Contents

Preface		vii
1	The management of creative disorder	1
2	Betting the company	23
3	The pursuit of critical mass	43
4	Big is necessary	62
5	The corporate challenge	91
6	Governments and innovation	109
7	The European Community policy for innovation	131
8	The multinational scene	145
9	Multinational enterprises and sovereign states	158
10	Managing the international trade game	178
11	Concluding remarks	198
Index		213

[handwritten annotation next to items 4–5: "my problems of size"]

Preface

The power of innovation – innovation pursued with massive technological and financial resources – was demonstrated across the whole gamut of the material aspects of human existence during the Second World War. For the first time in history, the entire scientific and technological establishment of the warring nations was marshalled to tackle not only the weaponry but the communications, supply and health of the whole community. The successes achieved, with cost only a secondary consideration to the principal aim of survival, demonstrated the ability and provided the existence theorem that, given enough intellectual and financial resources, almost any material problem of mankind could be overcome and any material want satisfied. To adapt Archimedes – 'Give me a big enough technological lever and I will change the earth.' Such innovation, far removed from the much slower evolutionary processes of previous eras, merits a distinctive name and we have chosen, for reasons which emerge in the book, to call it *macroinnovation*.

While the globe itself may not have been changed, the years following the Second World War saw it being dramatically modified as a result of this massive investment in innovation. Civil air transport girdled the earth, rocket-powered access to its outer limits became a reality, the power of the atom was harnessed and the quantum leap achieved in electronics in the War entered every aspect of everyday life.

In the victorious countries much of this was supervised and financed by the state machinery set up for the Second World War. Still heavily biased towards defence, it absorbed a significant part of the investment of industry in setting up and maintaining large research and development departments and so enabled many sectors to build up their innovative capabilities at the taxpayer's expense.

By the 1960s, however, this form of state-subsidized innovation of commercial products fell out of favour as its contribution to industrial competitiveness was, in general, manifestly less successful than the Japanese and German approach. Denied, as the vanquished, the continuity of Second World War type state intervention, these countries had had to develop alternative approaches which, while still involving the state, focused on market penetration rather than technological achievement. The progressive withdrawal of direct state financial support began with painful consequences as enterprises in the UK, USA and France felt the full costs and risks of competitive innovation in international markets for the first time.

From the late 1970s onwards, the result of this has been a progressive restructuring of the commanding heights of manufacturing industry on a global scale. Takeovers, mergers, joint ventures and the establishing of foreign subsidiaries abound, all with the objective of creating enterprises with the global market reach and commensurate financial strength required to match the costs of competitive macroinnovation. Even the defence industries, the last bastion of Government-financed innovation, are now regrouping into large transnational enterprises the better to respond to competitive procurement in this sector. State subsidizing of national champions has greatly reduced and state intervention to remedy what is perceived as unfair intervention in transnational competition increased. The activities of the Commission of the European Community (EC) tackling the dual problems of free trade within the EC and external relations with the USA, Japan and the Third World represent a major initiative in this respect.

Thus, both industry and sovereign states face new challenges; industry to establish and manage these new transnational structures and sovereign states to come to terms with their increasing dependence on multinational companies outside their national jurisdiction. National anti-trust regulation, national import quotas and national corporate taxation become increasingly ineffective and meaningless when enterprise on a global scale is the issue, when the miscegenation of its origin confuses the nationality of a motor car and when a true and verifiable breakdown of the profits of a transnational company, territory by territory, can become a minefield of political disagreement.

The purpose of this book is to examine this important development of a truly global economy and to consider its implications for both industry and government. The first part of the book – the first three chapters – examines the process of innovation in detail and explores its consequences for an enterprise seeking to compete in this market. From some simple assumptions, the inevitability of what is called 'the pursuit of critical mass' is explained. Only companies of a minimum size can face the costs and risks of entering a given competitive market; a market we characterize with its two main factors of selling price and sales volume. Cases where innovation amounted to 'betting the company' are examined and examples from a wide range of industrial sectors of the active pursuit of critical mass in recent years are given.

In the next two chapters we seek to put all this in the context of the overall management problems of an enterprise. Much has been written about the merits of smallness, the desirability of such an environment for individual initiatives to flourish and the bureaucratic dead hand of big organizations. Cooperative joint ventures which seek to meet the requirements of scale in financing and marketing innovative products offer an alternative route but are not without their own management problems. We examine these aspects, first under the heading 'Big is necessary' and second, in the wider subject of options for management, in 'The corporate challenge'.

We then turn to considering the vital role of the state in innovation. While the direct financing of innovation by the state is out of favour, and free market competition is avowed if not always observed, the state continues to have a considerable influence on both the innovative capabilities and the structure of enterprises operating within its boundaries. Subsidy, the allocation of taxpayers' money in support of industry, can be seen in the level of investment in technological education and research, in public purchasing, in regional development and offset deals as well as in the financing of state-owned enterprises. However much non-interventionists may dislike it, the aggregate of these activities constitutes a national policy. The coherence and effectiveness of such policies vary greatly from nation to nation. The advantages to be gained when such policies are derived from a consensus of industrialists and permanent officials, with the minimum of party political interference, emerge from consideration of the different national approaches.

The European Community has, as an element of its programme to create a single European market, followed the practices of its member states in its approach to innovation. In particular, it has launched a number of partially subsidized applied research and prototype technological development (R&TD) programmes in selected industrial sectors. Examples of these are examined and the case for them as a catalyst in developing trans-European attitudes is accepted. Their potential for damage, however, through industry becoming excessively dependent on such programmes for their innovative activities is identified.

At the same time, non-intervention or very limited intervention in innovation has the inevitable consequence that the commanding heights of industry will be occupied by multinational enterprises competing on a global scale. This globalization of industry is not being matched by a coherence of the approach of governments. Competition policy, taxation and many aspects of trade policy remain essentially national and guarded in the name of sovereignty. In the remaining chapters of the book we examine how this impacts on governments, on multinational companies and on the world economy.

This is the scenario in which the challenges to industry and the state must be faced. Industry to become global in outlook, at the commanding heights to be big and quasi-stateless and at the lower levels to become suppliers and subcontractors, regardless of the nationality of origin of the companies at the top of the hierarchy. The state, still accountable to its people for their economic wellbeing, must find new ways of intervening and regulating these enterprises, ways that must replace those we have described based on the concept of national industries, trading in a framework controlled by bilateral and multilateral agreements between governments.

The challenge to industry is to restructure itself to operate efficiently on a global basis while, at the same time, going as far as possible to be 'acceptable' industrial citizens of the disparate nations in which it operates. The challenge to sovereign states is to retain their role as protectors of their economic wellbeing while being 'acceptable' hosts of the global enterprises, who will increasingly occupy commanding positions in their industries.

We conclude by reciting the origins, importance and potential for good that properly regulated approaches to global competitive innovation can bring. Shining through all the arguments about the current Japanese imbalance is the need to diminish the 'winner takes all' consequences of a nation or company's temporary success in a market sector. To progress there must be winners but little need be lost if this winning process requires the benefits to be shared through transnational development. At the heart of the Third World problem is their lack of appropriate industrialization. This will not be achieved by money alone, which history repeatedly shows is often wasted through local inexperience, but needs the participation of suitably directed transnational enterprises. Finally, the great macroinnovative endeavours of space, of conservation and of world health can only be efficiently achieved through the proper use of transnational enterprise.

In this the EC initiative has a major part to play. If it can 'crack' the problem of developing a transnational approach in its own diverse sovereign states, this could be a model upon which to base the attack on the global problem.

The authors have brought to the task of writing this book two quite separate backgrounds of experience: one that of a senior executive with forty years' experience in most aspects of aerospace business management; the other that of a senior academic with many years' experience as an industrial consultant. Our purpose in this unusual combination was to balance the lifelong enthusiasm for innovative enterprise of the one with the breadth of knowledge and objectivity of a student of management of the other. Despite determined efforts on our part, it is not difficult to detect who wrote what! We hope that, despite any shortcomings that may be traced to this source, the ground we have been able to cover by this collaboration has made some contribution to illuminating this important subject.

Basil Blackwell
Samuel Eilon

1 The management of creative disorder

- Innovation's 'Big Bang'
- The linear model
- A market model
- Components of innovation
- Relative costs
- Cash
- Why bother?

Innovation's 'Big Bang'

Innovation is as old as mankind but for almost its entire history it has followed an evolutionary course in which natural selection rather than deliberate and systematic exploitation has determined its progress. Such technology as there was consisted largely of 'tricks of the trade' drawn from experience and only rarely recorded. The business of understanding natural phenomena existed in almost total separation from the world in which the craftsman lived and worked. Fernand Braudel (1981), in his history of the fifteenth to eighteenth centuries, documents at length how the great inventions in printing, in land and sea transportation and in the military use of gunpowder took centuries (even longer, if the seemingly inevitable Chinese input is taken into account) to evolve and become established in the everyday life of nations.

The first step change can be identified in the coming together of natural philosophy and craft skills in the coteries that grew up around the beginning of the nineteenth century. While natural philosophers have always sought applications of their knowledge, these needed to be designed in a practical form, made reliable and reproducible and sold at an acceptable price before they could be said to have been reduced to practice. To achieve this, a close partnership beween science and craft skill was essential, with mutual understanding and respect on both sides.

With this partnership came a much more pro-active attitude to innovation. The long-established products of craftsmen, like the steam engine, developed rapidly with the scientific understanding of the physical properties of steam. Such partnerships, epitomised by the close relationship between Watt and Black (of latent heat fame), played a major part in the innovations of the nineteenth century.

By 1831, the British Association had been formed and the products of the marriage of skill, science and private fortune were much in evidence in the 1851 Exhibition in London and the 1867 Exhibition in Paris. For the rest of the nineteenth century and the early twentieth century there followed a stream of inventions, amongst which were electric lighting, the internal combustion engine, telegraphy without wires and aerial locomotion. Throughout this period the emphasis was on the inventor and the monopoly, through patent protection, that provided the reward for the successful. Swan, Otto, Marconi and the Wright brothers appear not as industrialists but as individuals vigorously defending their rights to reward through royalties. Companies were established solely for the purpose of developing, making and selling specific inventions and the failure rate was high.

However, the survivors – and particularly those in the technology-based industries of power generation and the production of chemical and metallurgical products – started to build up their own 'in-house' innovative capabilities. Beginning with the improving and commercializing of the inventions of others, these increasingly started generating and patenting inventions in which the company, not the individual, held the rights. By the 1920s, the separate existence of the inventor and manufacturer had become blurred and many inventions were anonymous in the sense that, although made and patented by individuals, possibly employees of the company, they were developed, produced and marketed in the company's name.

The state has always been involved in innovation both in respect of its defensive (and offensive) responsibilities and in serving the public good. Public buildings, roads, bridges, water supply and the weapons of war have all been the subject of inventions which have, through state procurement processes, become innovations. To carry out this work, the state has had experts in the encouragement of innovation and in ensuring that such public works were properly executed.

With the coming together of science and craftsmanship, referred to above, the technological competence of the public organizations involved acquired an increasing level of importance. To be competent supervisors of state procurement they had to be at the forefront of technological development and, to achieve this, be themselves delving deeply into almost every aspect of the growing industrial technology base.

By the early part of the twentieth century there had emerged, in government service, establishments employing full-time research engineers and applied scientists. These developed into centres of excellence, rivalling the universities and attracting people of the highest calibre. In the UK, the National Physical Laboratory, the Road Research Laboratory, the Royal Aircraft Establishment, the Admiralty Research Laboratory and the Armament Research Department, and many others, had, by the late 1930s, succeeded in attracting a cadre of brilliant people (many of whom subsequently played leading parts in the innovative industries of the post-war years). Similar organizations had grown up in France, Germany and the United States.

Thus, when the Second World War started, a nucleus of people and organizations existed on which it was possible to build the massive technological effort employed by both sides in this conflict. The aeroplane changed in a few years from an interesting invention for the few to a fighting, load carrying and communication innovation for the many and, in the process, established the air transport industry of today. Radiolocation grew into the sophisticated radar and associated electronic systems activity from which a whole new industrial sector has now developed. The medical problems of survival in the jungles of South East Asia received an unprecedented concentration of talent, to which the post-war pharmaceutical industry owes much.

The systematic pursuit of innovation with which this book is concerned has its origins in this technological 'Big Bang' of the Second World War. Of course, examples can be found of it in certain industrial sectors before that time, but it is the unparalleled outburst of innovative activity and unprecedented conscription of talent of the Second World War that established it as a new way of life. The commercial and military approaches to innovation have developed in quite different directions in the ensuing years, but the existence theorem of modern competitive innovation – that 'almost' any requirement of mankind can be satisfied given enough financial and technical resources – was demonstrated by this wartime experience.

The linear model

An understanding of the process of systematic innovation in a competitive environment is essential to the study of its consequences for both industry and government. Because innovation is an omnibus description of the introduction of something new, and covers the whole range from supermarkets to space travel, we must begin by narrowing the field.

First, we are referring to innovation in the production of manufactured goods and their associated utilization, now generally known as software. We are concerned with the innovations that represent major changes in the way a need is satisfied, and not just incremental improvements in an established product. This has a most valuable part to play in extending the sales of existing products but does not present the order of challenge of an entirely new product with which this book is concerned. A challenge that if successfully executed can dominate a market and leave competitors struggling with a manifestly obsolete product – and if unsuccessfully executed can bankrupt a company.

Such innovation is both expensive to initiate and needs the greatest possible sales on which to recover the investment. Drawing deeply on existing technology it will probably require further research before a final product can be confidently produced. It will involve sustained effort over decades to launch and to achieve success through sales. All this, in our view, justifies the prefix *macro*, to distinguish *macroinnovation* from the much wider set of every type of newness which innovation covers.

Governments, the originators and financiers of many macroinnovations particularly in the defence sector, have historically found it necessary to present a veneer of orderliness in what is, inevitably, a disorderly sortie into unknown territory (if it were otherwise, it would not be innovation) (Office of the Minister for Science, 1961). The belief that, with good management, winners can be selected at the outset and marched in a predictable way through some ten years of development and production, has not been discouraged. After all, the reality of macroinnovation that we shall seek to present, of an uncertain gamble, is hardly what governments would seek to present as a proper use of public money.

Hitch and McKean (1960) in their classic work *The Economics of Defence in the Nuclear Age*, free of public accountability pressures, could afford to be more realistic:

> The important thing to appreciate in making good decisions with respect to research and development is the dominant role played by uncertainty...
> ... The military services have all too frequently tried to command the research and development community to invent new weapons to specification, just as they would command a platoon of infantry to march by the right flank.

It is to emphasize this very important point that we have entitled this chapter on the process of innovation 'The management of creative disorder'. In truth, it is a creative integration of the components of innovation from the technology base to the rapidly-changing nature of the competitive marketplace – a process more analogous to the way the brain makes sense of the confusing signals it receives than to a structure through which an established task is executed.

Nevertheless, this linear model of progression from science through technology to industrial success continues to be used extensively in official circles and we cannot proceed further without first devoting some attention to it. It all starts with the role of science in macroinnovation –

and with the mistaken idea that even science occupies a place in this linear model.

Figure 1.1 shows the set of sequential steps in this concept, starting from pure research and culminating in the production of goods and services to meet market demand. The participants in this process are the universities and industry with the former concentrating their effort on pure and applied research and the latter converting this research base into wealth-generating products.

Not only, as we shall see, is this a misleading model of the whole process but it leads to some potentially damaging conclusions.

With the general acceptance that national wellbeing is, in significant measure, dependent on participation in innovative industry, the demand for a simple criterion is considerable. 'Science', 'technology', 'R&D' (research and development) or even 'R&TD' (research and technological development, as the European Commission would have it) become politi-

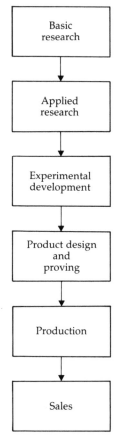

Figure 1.1 *The linear model*

cally important as activities that measure a nation's attention to this important wealth-creating activity. The pressure is then to direct public funds for research in universities and scientific institutions on the basis of their value as inputs to this linear model. The need for industry to 'commercialize' the output of this process is recognized but as a result of this 'technological push' rather than 'market pull'. When the R&D departments of industry are lured away from their vital role as part of the total innovation team of a company with offers of 50 per cent funding – and universities told that they must team up with industry if they want public funds for research – the model is in danger of becoming master rather than servant of policy.

There are, of course, great benefits from close relations between universities and industry and in so far as these policies contribute to this, they have a valuable aspect. The universities best serve the nation, however, through pursuing excellence in teaching and research with openness and publication of results. Companies, on the other hand, must seek competitive advantage through patents and commercial security. The university builds up the corpus of technological knowledge and skills of a nation, industry draws on this and uses a different sort of skill to commercialize it.

We turn in later chapters to examine more closely how governments and the European Commission view their role in this. However, not least to emphasize that other models have a claim to serious consideration, let us examine an approach to innovation from the standpoint of the market.

A market model

The higher the technology the more difficult it is to get the technical champion of a product to address the question of the market for it. Its newness is used to discredit any comparisons with previous products and even to imply that its outstanding technical merits will result in it being bought at any cost. Concorde, the supersonic civil airliner, was launched in almost total disregard of the economics of its purchase and subsequent operation. This attitude still survives in the recent so-called 'Supersonic Strasbourg' declaration of those devotees of supersonic transport technology, who seem to be dedicating themselves to studying a successor to Concorde without first establishing its economic feasibility for either manufacturer or operator (*Journal of the Royal Aeronautical Society*, 1990).

The difficulty of getting product pricing considered at the earliest possible stage of the innovative cycle is exceeded by the unwillingness to address the subject of sales volume – and yet it is only through sales volume that the recovery of the launch investment can be made and the all-important selling price constructed. But more importantly – and more subtly – the sales volume is almost always the master and not just the consequential outcome of successfully launching an innovative product.

Markets have characteristic selling prices and they have characteristic sales volumes. No matter how brilliantly designed and executed an innovative airliner which can only be acquired for ten times the price of the inferior alternatives, it will not sell. This is usually well understood. That an equally brilliantly designed and executed innovative airliner, whose price is comparable to its inferior alternatives but is only produced at the rate of two a year will not sell needs, surprisingly, some further explanation. There is a view that an innovative product can, as it were, creep into the market and build up its presence as a foundation for its ultimate success. That this is possible – the much publicized 'overnight millionaires' of Silicon Valley can be quoted – cannot be denied. However, the aggregated effect of these 'lucky breaks' on the totality of macroinnovation is minimal. The unfortunate generality is that the market must be attacked with the sales volume that is characteristic of it.

Planning a launch with a sales volume much greater than the market can absorb will, of course, be disastrous. Planning it for an order of magnitude less will greatly increase its chance of failure, because of a combination of adverse effects that limited availability brings, such as:

- Credibility (who else has bought it?)
- Imitation (the hiatus in supply lets inferior imitators in).
- Improvement (the hiatus gives competitors time to offer an improved version of the original idea).

The characteristic volume – V_c – of a market is not a matter of great accuracy. In market parlance, a penetration of as low as 5 per cent or as high as 35 per cent is the range so that a logarithmic scale, in which markets are quantified in multiples of ten, is adequate to the level of accuracy needed for our purpose here.

Typical examples of this 'categorizing' by sales volume are the following:

- $V_c = 10^0/yr$ – the product is being made and sold at the rate of about one per year, designed in direct contact with the user and largely unique to his requirement. The prototype is in fact the finished article. Examples: Satellite Launch Vehicle, nuclear power station, the Eurotunnel.
- $V_c = 10^2/yr$ – the product is being made and sold at about five per week. This will probably be a special product, designed so that it can be tailored in significant detail to suit each customer's individual requirements. Direct relations with each customer will be maintained through specification into after-sales service. Examples: aircraft, specialist machine tools, major electrical generating plant.

- $V_c = 10^4/yr$ – the product is being sold at the rate of tens of units per day. It will no longer be possible to handle this direct with the customer and sales will be through separate marketing and sales organizations working with comprehensive sales literature. This may well have significant technical content and customer satisfaction will be very dependent on the competence of the organizations to whom this is delegated. Examples: specialist farm machinery, luxury motor cars, TV studio equipment.
- $V_c = 10^6/yr$ – the product is being sold at the rate of thousands per day. This will probably be a mass consumer product, highly standardized and sold through a hierarchy of dealerships and retailers. Customer satisfaction will be totally dependent on the product's intrinsic quality and foolproof operating instructions. Examples: household appliances, popular motor cars, TV and video equipment, microcomputers.

Using market need (not technological research) as the genesis of innovation, and with the understanding of the sales volume and price implications of the target market, an entirely different model of the innovative process emerges. This is shown in Figure 1.2.

The product conception could be a camcorder with twice the capabilities and half the price of its competitors; or it could be a cure for AIDS. The point is that the product conception or conceptual invention comes from an evaluation of the market and not from the 'backroom' of research. Furthermore, the translation of conception to a defined product uses the most economical and expeditious mix of buying-in (by licence for example) and in-house technology. Following product definition, the two models proceed along identical routes, but with a much higher probability that the product is optimized for producing and selling than is the case in the linear model.

Components of innovation

The truth of the matter is that neither the linear nor the market model of innovation is anything but an oversimplified version of the disorderly reality of a creative activity. In Chapter 5, we present yet another diagrammatic representation of the process of product design – a veritable 'spaghetti junction' of mutually overlapping and conflicting streams of ideas out of which a hopefully optimum design will emerge.

Nevertheless, while the interactions are complex, it is useful to break the totality of a launch investment into a number of discrete components.

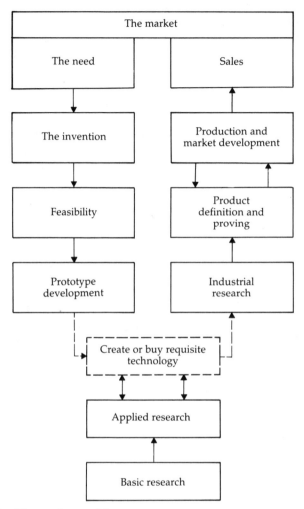

Figure 1.2 *The market model*

These are:

- Invention and feasibility
- Product definition
- Production launch
- Marketing and product improvement

In all these, the totality of the problem is present. Marketing in the feasibility stage, production during product definition and the potential for subsequent product improvement in the production planning. They are best viewed not as isolated subdivisions of the professional skills involved but as convenient milestones in the build-up of the total launch investment.

Invention and feasibility

The quintessential and yet least costly element in the innovative process is the idea, the invention. Despite it being extravagant in the sense that only '1 in 100' passes the feasibility test, the cost of providing the environment in which minds perceive new relationships between knowledge and need is not the issue – it is in providing them with the conviction that the resources will be made available, if the idea passes the test of feasibility, to realize it in marketable form.

The foundations on which feasibility is established are of supreme importance in competitive innovation. The market and the timing of entry are, as we have indicated above, also important and a realistic assessment of the total launch cost is vital. As we shall see later, the availability of adequate financial resources and the capacity to face the risks involved determine the appropriateness of a company to launch an innovative product, regardless of its brilliance and sales potential.

However, the more advanced the technology involved, the more important it is to establish beyond reasonable doubt that the technical specification of the invention is achievable. This is the realm of the prototype demonstrator, or the so-called 'bread board' mock-up, and of applied industrial research to fill the gaps in the technology required to go forward with reasonable confidence to the full design stage.

The amount of work involved and, therefore, the cost of this aspect of the feasibility component varies enormously with the type of product and the technical advance involved. An extreme example is provided by the approach to manned supersonic flight in the 1950s and the even more complex establishment of the feasibility of manned space flight in the 1960s. In both cases very large investments were made at the feasibility stage culminating, in the case of supersonic flight, with the very public drama of the work of one USAF pilot, Chuck Yeager, who literally rocketed himself through the sound barrier in the Bell X-1 test vehicle – it was in no sense a usable aeroplane. These tests provided an essential pointer to the further research necessary to make supersonic flight a feasible proposition. While less dramatic in popular terms, most feasibility programmes feature such tests deliberately, designed to focus the supporting research on the ignorance that must be remedied before an invention can be considered technically feasible.

To carry out this work, a special type of organization is required. One such which has achieved world renown in the aeronautical sector was founded by Kelly Johnson of Lockheed in 1943. His 'Skunk Works' unit – more properly, the Advanced Development Projects unit of the Lockheed Company – acquired its name from comparison with the mysterious skunk processing factory in Al Capp's comic strip *Li'l Abner*. The management philosophy of this unit, which launched the United States into the jet propulsion age and later demonstrated the feasibility of the secret reconnaissance aircraft at extreme altitudes (the U-2) and extreme speeds (the SR71 Blackbird), was described by Rich in the 1988 Wilbur and Orville Wright Memorial Lecture of the Royal Aeronautical Society (Rich, 1989).

Finally, there is the all-important consideration of product cost. As a concept takes on a physical form – and before it proceeds to the detailed design of product definition – its compatibility with the price characteristic of its intended market must be tested. Starting from a simple 'parts count', in which the number of parts is compared with those in the nearest comparable product of the past, this can be developed into the more sophisticated method of product cost estimation in which each part in the parts count is ascribed a cost weighting. This process of parametric costing is not intended to do more than reject inventions whose prices would be outside the characteristic price level that the target market will stand. Designing for a competitive price occurs in the production definition element of the launch process.

With all the economic and technical inputs in place, it is possible to construct a business case. This is then checked for its feasibility, ranked alongside other innovations competing for the finance available and against the alternative of deferring or 'doing nothing' – with the short-term benefits and long-term risks to the company that this entails.

If it passes all these tests, the innovation then passes to the more expensive component of product definition.

Product definition

In the search for a name to identify this component of innovation, *product definition* seemed to contain a greater generality than design and development, product engineering, standardization for production, etc., all of which are used in industry. All these activities are directed to defining in exact and unequivocal terms what is to be marketed.

To do this requires a design programme whose end product is a physical specification, accompanied by a test programme proving that specification and ensuring that it complies with the requirements of the appropriate independent regulatory authorities.

However, the product cannot be defined in isolation from considerations of production and marketing. In parallel with the definition process and with a strong feedback into it, studies will be undertaken on how it is to be produced and marketed. Both are akin to the feasibility studies of the product referred to above, aiming to establish detailed targets for cost, delivery date and sales volume and the investments required, while leaving the precise means of achieving them to be settled later. This division is technically artificial but financially very important if, as is usual, the decision to launch production will not be taken until a second business case has been made. This will have the benefit of the much more detailed information now available both internally and in relation to the competition. It will be exceptional if the undesirability or even impossibility of realizing all the technical and commercial aspects of the earlier feasibility case have not emerged. Relaxation in some areas, improvements in others and changes to deal with an updated view of the market will have arisen. All this has to be built in and the resulting business case subjected to critical review.

The importance of this component of the innovation process in high technology products cannot be overemphasized. It is a complex interdisciplinary activity and demands very special management skills of the highest order. In a paper entitled 'Design in the Management of Technology' Coplin (1986) of Rolls-Royce gave a lucid description of what is involved in the aero-engine industry. In the summary at the end of his paper he says, 'Design is the process concerned with packaging technology, blended with experience to match the market pull, with a product that can be underwritten by the technology and matched to the manufacturing process.'

However, the decision to proceed to the next phase is not a foregone conclusion. Terminating, or possibly referring back on timing or market considerations, may well be judged the proper decision. The point has already been made that picking 'winners' at the outset and then marching them through the innovative process is not the formula for successful innovation. Stopping or deferring at this stage is not to be considered an inexcusable failure of the process. As we shall see in Chapter 2 innovation is a high-risk business in which an allowance for failures must be made. Stopping at this stage is far less costly and damaging than going ahead and experiencing a failure in the marketplace.

However, if it is decided to go ahead, the next two components of investment in innovation, production and market planning, will then be embarked upon.

Production launch

The production component of the innovation investment also varies greatly with the nature of the product. In many, it embraces the all-important make or buy decision, the selection and involvement of proprietary equipment designers and manufacturers with the opportunity to off-load part of the costs and risks of innovation this presents.

Through the product definition activity, production planning of the emerging product will have proceeded and the targets for cost, delivery rates etc. have been established. Research into innovative ways of improving the production process will have been launched and the costs of the new equipment required, the new factory layouts and any new methods of production control will all have been fed into the business case.

It is often not recognized that a vital element in the launch cost of a new product is the build-up of inventory before sales commence and the non-recurring losses which have to be recovered in sales due to the time it takes to get production down to its target costs. The former can be reduced by ingenuity in the information systems of production and subcontractor control and the latter by investment in manufacturing control in which the traditional 'learning curve' is minimized and early batches come out as near to the target cost as possible.

Finally, investment in quality assurance is a crucial decision in production planning. Nothing can be more expensive and damaging in the launch of an innovative product than the emergence of a quality defect in the full glare of the marketplace. While it must be assumed that functional problems will arise in service – possibly due to causes that were not foreseeable – subsequent investigation and remedy will first focus on quality. If the quality control records are impeccable and the remedy or replacement parts are available very quickly, the damage can be minimized. Production must be planned to take this important aspect into full account.

Marketing and product improvement

The plan for marketing a new product can involve considerable non-recurring investment. This is often overlooked in the R&D minded approach to innovation. It falls into two parts:

- The plan to promote the targeted sales.
- The plan for the technical support of the sales achieved.

In the first part the numbers, skills and geographical spread of the personnel required must be determined, their training initiated and the aids such as demonstrations, advertising and appropriate literature designed. Some form of test marketing will probably be required and investment in special financial assistance to first customers allowed for.

In the second part, a corresponding review of personnel to provide customer support must be planned and implemented. A customer training programme, test equipment and diagnostic aids and spares support will be required and be promised at the point of sale. Machinery for feeding back customer problems and getting them dealt with expeditiously must be set up.

Where dealer networks are required, the marketing plan must be tailored to build up their sales and after-sales capabilities in these respects.

The innovative process rarely ends with the sales of the product originally launched. In the worst case, non-competitive features may emerge in the initial promotions or in early sales, which make it a matter of urgency to produce an 'improved' version as fast as possible and withdraw or exchange the original product. However, even if the original product achieves acceptance, the feedback from experience in use and the competition will make it necessary to face the decision (including the timing) of an 'improved' version. The business case will be easier to prepare and the investment less risky than that for the original launch. It will, however, involve significant investment and this will probably have to be made quite early in the sales programme – adding to the outflow of funds, and so needing some provision in the total funds required.

Relative costs

In the 1960s, the United States Secretary for Commerce established a panel of distinguished industrial and academic experts to explore ways of improving the climate for technological change in the US. This ad hoc 'Panel on Invention and Innovation' reported in 1967 under the title *Technological Innovation: Its Environment and Management* (Panel on Invention and Innovation, 1967).

In a chapter entitled 'Innovation in context' it sought statistical evidence on the costs of innovation and found that:

> Such data as are available primarily concern research and development, not the total innovative process, of which R&D is only a part. These data give us a reasonable indication of the investment in R&D, who is performing it and to what extent. But they are not reliable indications of innovative performance.

Finding it impossible, however, to proceed with their task of advising the US Government in this situation, the panel resorted to pooling their collective, personal knowledge and found 'that there was sufficient similarity in the experiences we covered to convince us that it would be desirable to present these rule-of-thumb figures as a basis for discussion.'

Their figures, recast under the four headings we have chosen to define the elements of launch costs, are as follows:

- Invention and feasibility   5–10%   (Av. 7½%)
- Product definition   10–20%   (Av. 15%)
- Production launch   40–60%   (Av. 50%)
- Marketing and improvement   15–40%   (Av. 27½%)

These figures, presumably drawn from experience in the wholly commercially-financed innovation process of the 1960s, probably relate to the higher volume consumer-type market with sales volumes of 10^4 and above. At the other extreme, we can derive a rough estimate of the preparations for a lower volume, high technology product from the Rolls-Royce prospectus issued in 1987 by the UK Government when its shares were publicly offered (Rolls-Royce, 1987). From these it is possible to guess that for a new aero-engine type, the following division of cost might apply:

- Invention and feasibility   15%
- Product definition   50%
- Product launch   20%
- Marketing and development   15%

with the bulk of the 'marketing and improvement' investment going into the in-service development of overhaul life. The development of derivative products, a universal practice in extending the sales of a new aero-engine type, will involve a separate further investment outside this breakdown.

Published statements in the pharmaceutical industry indicate an equally high proportion of the launch costs going into product definition, both having stringent safety requirements involving exhaustive testing to the satisfaction of independent regulatory bodies before they can be sold. In the case of the pharmaceutical industry, the further investment in market development appears to be exceptionally high. Published figures indicate as large a level of employment of professional staff in the marketing element as in product definition.

With all this imprecision, one conclusion can still be drawn. This is that in terms of launch investment, the R&D element is by no means the whole story – and in many cases it is only a minor part of the total cost. This is illustrated in Figure 1.3, a 'trend curve' which can be postulated to illustrate the rapidly declining proportion of 'launch cost' which is 'arandee' as characteristic sales volume increases. Using a generous definition of 'arandee' to include the whole of product definition, even at the low volume end of the aircraft industry only some two-thirds of the launch cost is 'arandee'; at the high volume end of the automobile and consumer electronics industries it could be as low as a fifth or less. This is important when considering the value of the various forms of government aid for innovative industry – in financial terms it may only be a partial contribution to the tip of an iceberg of investment that industry must make to achieve a marketable product.

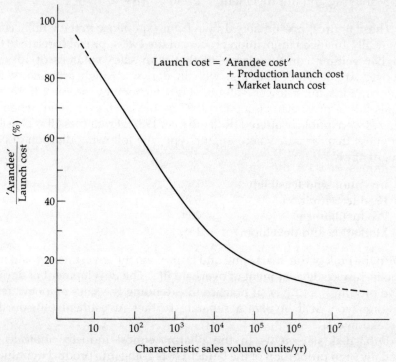

Figure 1.3 *The proportion of launch cost that is 'arandee'*

Cash

The cumulative cash flow profile

We now turn to the subject of cash, the hard and unyielding master of all enterprises and particularly important in the confusing disorder of innovation. As every inventor knows, having enough to launch a new product determines whether it sees the light of day. As every inventor finds out, sometimes very painfully, generating enough in subsequent sales to repay (with appropriate interest) this input of cash is essential for commercial success.

This issue may be clouded in established enterprises. With a variety of products at various stages of maturity and obsolescence, and by treating some of the costs as revenue and capitalizing (i.e. assigning a future value regardless of the success of the product) others, the issue can be fudged. However, all these things need to be paid for and when the borrowing requirements of the company exceed the facilities the market is willing to make available, cash returns as the master of even the largest and most diverse of enterprises. Running out of cash is, as we illustrate in Chapter 2, the greatest single cause of failure in innovative enterprises and access to enough of it is essential to success.

The cumulative cash flow profile of an innovative product gets to the heart of this matter. A typical profile of a successful product is illustrated in Figure 1.4. It begins with an outflow of cash as expenditure on the various stages of the innovative process described above proceeds, slows down and turns round as sales commence and a surplus of income over cost of sales develops. This may not occur initially since early batches may cost more than the selling price. As sales develop, this cash inflow continues until, if the product is a success, the break-even point is reached and net cash generation begins. This process continues until the product becomes obsolete and sales finally cease.

This evolution of sales, with product improvements sustaining them and deferring obsolescence, is itself a complicated process which is discussed further in Chapter 5. Its effects are secondary to the first order treatment we are presenting at this stage.

In Figure 1.4 the critical dimensions of this profile are shown. These are L, the maximum negative cash flow which we shall call the launch cost, Y the number of years from initiation of launch to this point and nY the subsequent lapse of time before this negative cash flow L is recovered and the product starts to generate a surplus over that required to launch it.

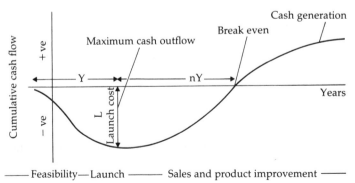

Figure 1.4 *Cumulative cash flow profile of a typical successful product launch*

In the macroinnovative products with which we are concerned here, Y seems to lie between five and ten years – even longer in major advances and possibly shorter in the frenetic innovative periods in consumer electronics. The multiplier n varies with the dominance planned and achieved by the product in the market, but $n = 2$ seems a typical figure for an average success story.

The importance of the cumulative cash flow profile in the conduct of an innovative enterprise is, regrettably, not matched by the availability of information on it in company reports. Indeed, it could be said that the annual reporting system is almost irrelevant in presenting 'a true and fair view' of a company's innovative activities. With progress on a specific innovation measured in a timescale of decades, the relative 'snapshots' of the annual reports contain a summation of the past and current individual innovative cycles and give a distorted assessment of the future value of present innovation. It is distorted because of the conventions regarding what can be capitalized and what cannot – know-how and the technology invested in a new product being written off as incurred whereas the tooling, machine tools and stocks and work-in-progress building up to a product launch are capitalized.

This, and the fact that financial markets are based on present values of an enterprise, have far-reaching consequences to which we will return in later chapters. For the present, we can only regret that the justification of the order of magnitude assessment we present in what follows depends on figures solely derived from personal experience and hearsay.

The scale of cash outflow

Launch cost defined as the maximum outflow of cash in the cycle of launching a new product results in a larger, in many industrial sectors much larger, figure than that normally attributed to R&D expenditure. This may only cover the invention, feasibility and product definition aspects of the launch process and be only a half, or much less, of sources of cash outflow. Not only are there the further costs of production and marketing launch, but the extra costs arising from the build up of materials and components before sales begin, the production learning curve and the cost implications of correcting technical deficiencies in initial service.

Approached from this standpoint, it is impossible to make the order of magnitude assessments of launch costs needed to understand their implications for the scale of the enterprise involved. However, if the problem is approached from the standpoint of achieving commercial success in the marketplace, a much simpler, if admittedly cruder, assessment of launch cost can be derived. We have seen that at the outset and throughout the execution of a new product launch, it is essential that it is predicated on a selling price P and a sales volume V (units/yr). Furthermore, we have introduced the concept that the market into which this product will be sold must be identified and that it will have a characteristic selling price P_c and a characteristic sales volume V_c.

We need two further simplifying assumptions. The first is that a constant fraction k of sales represents the excess of cash generation over that used in manufacturing the product. The second is that the sales volume and selling price is constant from start of sales to the break-even point (in cash flow terms).

Then using the notation of Figure 1.5:

L = k × nY × PV
if k = 0.1, nY = 10 years, then L = PV
if k = 0.3, nY = 3 years, then L = 0.9 PV

could be two extremes of, say, long cycle aircraft manufacture and novelty consumer electronics production. Other combinations of k and nY will, of course, give different relations between L and PV, but, we suggest, the variation is not that large.

We, therefore, propose to postulate that, in order of magnitude terms the scale of total investment required to execute a planned new product launch into a market with a characteristic price P_c and a characteristic volume V_c is given by

$L = P_c V_c$

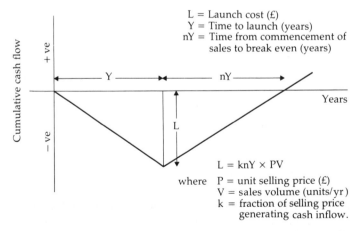

Figure 1.5 *Launch cost estimation from a simplified cash flow profile of a successful product*

This can be related for the following disparate innovative products, the total launch costs being shown below:

	P_c	V_c	$L\ (=P_cV_c)$
Camcorder	£750	10^6	£0.75bn
Popular motor car	£5000	5×10^5	£2.5bn
Main frame computer	£2m	10^3	£2bn
Intercontinental airliner	£30m	10^2	£3bn
Space station	£20bn	1	£20bn

The limitations of the pseudo-numeracy of this formula are, of course, considerable. The qualitative benefits, however, in terms of encapsulating principles that would take many words to express justify, in our view, its usefulness in assessing the financial requirements of innovation. These are:

1 That, other things being equal, launch cost is proportional to product selling price.
2 That, equally, it is proportional to planned sales volume.
3 That, despite all the complex elements of R&D that make up parts of the launch cost, it is the market criteria of P_c and V_c that dominate the launch investment.

We hasten, of course, to add that in any specific launch, skills in cutting down L by good management are vital and most worthwhile. These, however, are never likely to alter the outcome by an order of magnitude. It is to this level of accuracy that the formula aspires and from which we will later derive the scale of enterprise (company, joint venture etc.) that can prudently launch an innovative product in a competitive market with characteristics P_c and V_c.

Why bother?

We have sought to establish the issues that an enterprise embarking on what we have termed *macroinnovation* must face. They are extremely complex, intensely interactive, expensive in both the quality of intellectual and managerial skill and in the 'patient' financial investment required. As we shall see in Chapter 2, the risks of failure are far from remote. Why bother?

The central problem of manufacturing industry has always been that its products have a finite life cycle. The advent of the present macroinnovative age has reduced these life cycles enormously. If this were not enough, it has also opened up the possibility that, through macroinnovation, whole industries may be made obsolete – or, at least, reduced to a tiny fraction of their former size.

This technology-based, market-driven race is now so competitive and so fierce that companies – and, indeed, as we shall see later, nations – cannot afford to be left behind.

There are, of course, underneath this frenetic activity, the industrial commodity markets where the 'nuts and bolts', the 'connectors and switches' that are common to a wide range of innovative products are made. In his 1985 lecture 'Earning an Industrial Living' Alan Lord put it this way:

> Unless one is prepared to accept the extremely uncomfortable disciplines of seeking a living – and it may well be a precarious one – in a commodity market for manufactured goods . . . there is no practical alternative to driving the product range further and further up-market.

Uncomfortable – because the commodity manufacturer has no control and only minimum visibility of what will be demanded of him in the future. He may be the most efficient widget maker, but the demand for widgets is controlled by those above him. Precarious – because his competitive skills are restricted to the management of capital and labour, both wide open to competition from subsidized capital and cheap labour in the lesser developed parts of the world. But above all, quite inappropriate to an established industrialized nation whose assets lie in knowledge and whose liabilities are a public who expect standards of living comparable to the best available in the world.

The consequences of meeting this challenge on a global basis, however, are considerable. In Chapter 2 we look at what, for companies, can be the most dramatic outcome, the risk of failure.

References

Braudel, F. (1981), *Civilisation and Capitalism 15th–18th Century*, vol. 1 (translation from French), Collins.

Coplin, J. F. (1986), 'The Role of Design in the Management of Technology', *Int. J. Technology Management*, vol. 1, nos 3/4.

Hitch, C. J. and McKean R. N. (1960), *The Economics of Defence in the Nuclear Age*, Harvard University Press.

Journal of the Royal Aeronautical Society (1990), 'Supersonic Strasbourg', *J. Royal Aeronautical Society*, Jan.

Lord, A. (1986), 'Earning an Industrial Living,' *J. Policy Studies Institute*, vol. 7, part 1.

Office of the Minister for Science (1961), *Report of the Committee on the Management and Control of Research and Development*, HMSO.

Panel on Invention and Innovation (1967), *Technological Innovation – Its Environment and Management*, US Government Printing Office.

Rich, B. R. (1989), 'The Skunk Works Management Style – 77th Wilbur and Orville Wright Memorial Lecture,' *J. Royal Aeronautical Society*, March.

Rolls-Royce (1987), Offer for Sale on behalf of Secretary of State for Trade and Industry – Samuel Montagu & Co., London.

2 Betting the company

- The odds
- Hedging – the product range factor
- Credit and credibility – the financial market factor
- The civil aircraft industry
- The concept of critical mass

The odds

We have seen that to compete in the macroinnovative market requires massive investment, long pay-back periods and sales outlets comparable to one's competitors. It also demands the financial strength to absorb 'failure' without hazarding the very continued existence of the enterprise.

This risk recognition does not come easily to those on whom the driving force of macroinnovation depends. On the one hand, the enthusiasm of its 'champions' – the protagonists in the company without whom the innovation would not exist – coupled with some past record of success can encourage the belief that failure need not be catered for.

On the other hand, risk recognition can be so dominant that ever more sophisticated techniques such as market studies inches thick and exhaustive feasibility studies can produce 'paralysis by analysis' and the market is missed altogether. Continued risk evasion in an innovative industrial sector will itself guarantee the failure of the enterprise in time.

When those involved in the business of innovation can be persuaded to talk, away from the need to breathe confidence into shareholders and financial institutions in respect of their companies, some insight, albeit anecdotal, into the magnitude of the risks involved is revealed. Such an insight was provided in a series of BBC programmes, presented by Mary Goldring, in 1986 (BBC, 1986). From a panel of top people from industry and venture capital, she skilfully extracted the following views and opinions relevant to the assessment of the costs and risks of innovation:

- *Lord Keith (on Beechams)* 'Well, you do have instances where you have had bad luck, where you get a drug to a certain stage and then, in the final stages it doesn't stand up or produces some undesirable side-effect and it has to be scrapped and you have to start again, and it will

have consumed a good deal of money and, what is equally important, a good deal of time.'
- *Sir Denys Henderson (ICI)* In answer to her question 'What happens when absolutely nothing comes out of the labs and an awful lot of money is going into them?' – 'Well, firstly you pray hard. But we have been fortunate in that we have had a number of really dramatically successful products Inderal and Tenormin both of which are cardiovascular drugs, they have kept us going very nicely.' In answer to 'How long have they taken to develop?' – 'Oh, a drug these days takes the best part of twelve years, and costs an awful lot of money – anything up to a hundred million pounds per product. So you have to be successful in this game. This is a game for the big boys.'
- *Sir Ian Morrow* In answer to 'What proportion of companies would you say get into trouble because their products are not evolving?' – 'The rule of thumb is how long can we keep improving the present products? – which you can. And at a certain point the market is going to be ripe for new developments . . . If you come too early, it's a disaster, if you come too late it's a disaster. It's a very, very fine piece of judgement.'
- *Mr Brian Manley (Philips)* In response to 'Behind every compact disc is a laser, demanding and getting the sort of technology that you need for Star Wars. Now compact discs were the invention of the Dutch Philips company. Brian Manley is their research director here in Britain. It is received wisdom that such expensive pioneering does not pay.' – 'That received wisdom was true in the past and, perhaps, one can still find examples of where the trailblazer gets it wrong. The power of being first we see right now will be the compact disc. The ultimate is for a new product to achieve a penetration of 5 per cent of the world market in five years. That has actually never been attained. It now looks as if, in a period of three years, the compact disc will achieve that 5 per cent penetration factor.'
- *Mr Brian Manley* In answer to 'It takes huge research spend over decades to get such technical dominance – Philips spends almost as much on research as the British universities combined.' – 'Worldwide we employ about 4000 people in research and I would say something like 10–20 per cent is blue sky. Our total bill for research and development together is some 800, 900 million pounds.'
- *Lord Keith (Rolls-Royce)* 'The life cycle of a jet engine is anything up to thirty years. The Rolls-Royce RB 211, the first ten to fourteen years, the development phase, is almost continuous spend. I don't suppose you could design and build an engine from scratch for under two billion pounds.'
- *Mr Derek Roberts (GEC)* In answer to 'When does the big money have to be spent?' – 'I suppose the best single example of this is in the field of

silicon-integrated circuits. Now a major new facility is about 100 million pounds, that's the cost of land, buildings, all of the super-clean facilities and of the equipment that has to go into it and the initial start-up cost.'
- *Mr Brian Long (Acorn Computers)* 'Acorn was set up in the late 1970s and was really just a very small company until it signed the agreement with the BBC under which a microcomputer would be sold which was called the BBC Microcomputer. The company then grew dramatically to approximately 80 million in sales in 1984.'
- *Mr Brian Wilkinson (Dixons Ltd – retailers)* In answer to 'What went wrong?' – 'The market was hyped up – manufacturers overproduced 500,000 – 600,000 pieces too many for the demand . . . there was a great surplus of semiconductors. People could pick up computers at very low prices . . . prices came from 180 pounds to 130 pounds overnight. At the beginning of 1985 the distribution trade was very upset and the consumers were a little turned off.'
- *Mr Brian Long (Acorn Computers)* In answer to the statement 'For Acorn it was a tragedy. That it didn't go out of business completely is due to Italian Olivetti. Brian Long is now rebuilding Acorn, but at half its original size.' – 'The kind of market we are now pursuing is a much harder slog. Schools nowadays are much more sophisticated purchasers of computers than they used to be . . . We are trying to be less involved in that whim market.' In answer to 'Do you reckon a small, fast-growing company based on one product does need some kind of Big Daddy?' – 'It will eventually if it sticks as a one-product company.'
- *Mr John Ingledon (venture capitalist)* In answer to 'I have always understood that small companies are born and die like fruit flies.' – 'Out of five investments, three go bust, one doesn't go anywhere – that incidentally is the one that drives you bananas because it's still there – you still have to keep reading the papers and going to meetings and it's always about to come right, and one succeeds ten times over. And so with a little luck you double your money, you hope to do that in three or four years and statistically that pattern tends to recur and the pattern of new product development is about the same.'
- *Mr John Moulton (Schroders plc)* 'Out of the 100 or so propositions that come through the door we probably only do one or two. Out of the deals we actually back I would hope to be seeing between five and seven moderately successful to very successful, a couple which form the unattractive category of the living dead, companies that just trundle along happily having found their way into a very small market and probably one or two out of the ten who will actually go bust.'
- *Mr John Ingledon (on the Hilger Instrument business bought from the Rank Group)* In answer to 'Tell me a bad year – a really horrible year you would rather forget.' – '1983. We decided we should be in the X-ray

instrument business again and we simply got it wrong. In proportion to what was spent that year, it must have been of the order of a quarter, possibly a third. It's very difficult to say because if one thing is certain, nobody is going to admit that what they did spend is as large as it was. On a successful venture everybody claims credit and it is amazing how easy it is to find the exact expenditure. On the product development that is a failure no one was to blame and it didn't cost very much.'

And in answer to the question 'Does that happen to large companies?' the following was extracted:

- *Sir Denys Henderson (ICI)*   'With all the worries there were about smoking and cancer, we did actually produce a product that would have the same flavour as tobacco – and it would be inherently safer. The only trouble was that people did not like it.'
- *Mr Derek Roberts (GEC)*   'There was one particular area of research on a flat panel display technology, based on electroluminescence. And I'd always felt this was a good piece of technology. The trouble was that it was really in danger of being by-passed by other approaches to flat panel displays – particularly liquid crystal technology. And I believe that I was slow in stopping that work. I did stop it, but I think I should have stopped it a year earlier.'
- *Sir Kenneth Corfield (ex Chairman STC)*   Answering questions on how much optical fibre development cost. – 'The gestation period was fifteen years. If I were to put it at today's money terms then I would think we are talking about 200 million pounds.' And in answering a supplementary question, 'How many projects can you run at any one time?' – 'Depends on your cash flow of course. It's a function of your size; not the size you are going to be but the size you are at the time. Now when we at STC developed the optical fibre business and also some other things simultaneously we, it must be remembered, belonged to ITT. We were 100 per cent owned by ITT and, although we had a great deal of autonomy, we had the fall-back situation that if something went wrong you had to explain what went wrong to a perhaps impatient but very understanding investor. I don't think STC at that size and as an independent company would necessarily have done that or been able to do it.' Followed by the comment 'Ken Corfield's troubles date from this point. Without Big Daddy, his actions, his mistakes were measured against a different set of standards. The standards of the professional banker and fund manager.'

This admittedly anecdotal evidence reveals a general acceptance of the high risks in innovation. Failure, in the sense that the investment made is largely lost, can take many forms but three can be selected as representative. These are shown in Figure 2.1:

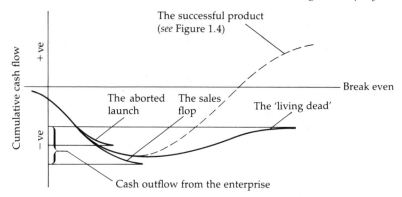

Figure 2.1 *Cumulative cash flow of an innovative product. Three typical futures, showing resulting cash outflow*

- The abandoning of the launch of a new product (aborted launch).
- The failure of sales after the product has been launched (sales flop).
- The achievement of only a fraction of the sales required to recover the launch cost (the 'living dead').

In Chapter 1 we presented the cumulative cash flow profile of a commercially successful product launch. This is reproduced in Figure 2.1 with the impact on cash outflow of the three 'failure' cases superimposed. In each case, to the accuracy of an order of magnitude, it is clear that the enterprise faces a cash outflow of the order of the planned launch cost L.

The probability of this loss being incurred can also be assigned in order of magnitude terms. The venture capitalists put this at a figure, in order of magnitude terms, of a one in six chance. Behind the guarded comments of the chairmen of large innovative enterprises a not dissimilar figure can be detected. One in six of course has a particular attraction as it corresponds to the chances in rolling a dice – or in playing Russian roulette! It is a good gambling number – not so adverse as to be not worth risking, but yet risky enough to be called gambling.

We, therefore, propose to model the risk of launching a new product as equivalent to betting an amount L (the estimated total launch cost) of the cash resources available to the company with a chance of losing it in one out of six cases.

Hedging – the product range factor

The ability to absorb such a loss is related to the range of products (and the stage in their evolution they have reached) from which the total cash-generating capability of the company is derived. Four types of business serve to illustrate the different degrees of exposure in which an enterprise can be placed in this respect:

- Case A – a single product start-up business.
- Case B – a single product follow-on business.
- Case C – a few products business.
- Case D – a multiproduct business.

In case A, the whole of the resources available to the company are risked and there can be no question but that the future of the company has been bet on the outcome. Investment in such a business is only financially prudent if the investor holds a portfolio of business in which this potential loss is small relative to its aggregate value. This is, of course, the classic venture capital case and is well recognized as such.

Case B, not so well recognized, is that of the successful case A company that must find a follow-on new product. Because of the shape of the cumulative cash-flow profile of innovation, and the lead time involved, this successor must be launched before the 'sunny uplands' of recovery of the original investment have been reached. This second injection of capital may well need to be followed by a third for yet another product before the business is able to finance new products from its internal cash-generating ability. This is based on the rough rule that an innovative enterprise can only be self-financing if it has one product range being launched, another in the full flood of cash generation and a third approaching maturity. If, on the way to this stage, say, the second product is a failure, then the all too familiar fall from grace of a 'golden boy' of innovative enterprise will be enacted. Dramatic rescues by larger organizations with deep enough financial pockets – or a major financial and management reconstruction – will probably be required.

Case C companies, while not exposed to the hazards of cases A and B, will experience a 'roller-coaster' fluctuation in profits which, while not bringing them to insolvency, will expose them to adverse publicity, restrictions on raising further money when it is most needed and even, in certain corporate financial situations, to a hostile takeover bid as their shares drop in price.

Case D is, from the standpoint of innovative risk, the happiest position for a company to be in. The impact of a one in six failure rate will, for them, be no greater than the many other hazards with which they are faced and have to deal in the normal course of business.

Credit and credibility – the financial market factor

The attitude of the financial market to an enterprise – innovative or not –

has a major effect on what risks a company can get away with. Lose the confidence of those on whom it depends for money, and the survival of the company is in jeopardy.

The market depends for its information on the company's annual report and accounts, supplemented by the press releases and presentations it may make throughout the year. Let us follow the impact of an innovative product on the annual report and accounts as it makes its way, over the years, from conception into success in the marketplace. It starts out by reducing the profit and increasing the capital employed as the elements of the launch cost are either treated as R&D and written off against profit in the year in which they are incurred, or capitalized and appear as increased depreciation. As increased borrowing is also likely, interest charges will increase and further reduce the 'bottom line' presented to the market. As the completion of the launch phase is approached, an increase in stocks and work-in-progress when production starts up will further increase the capital employed and the consequent increased interest charge to be deducted from the operating profit.

The best the board of the company can do to counter this black picture of the company's results is to include some words in the Chairman's Statement about its optimism for the future of the product in question! Furthermore, this adverse effect can extend over several years until peak launch cost is passed and sales develop.

What happens as a result of this depends critically on the nature and culture of the financial market concerned. The principal distinctive features are the structure of the stockholding, their powers in respect of the management of the enterprise and, through the rules governing trading of the stock, their powers and incentives to sell the company. In the UK and USA, for example, the stock of a company is usually widely held with no owners of dominant significance and is actively traded in a highly developed stock market. As a result, the price of the stock is not only very volatile in relation to the trading results of the company – and to the prospects for the sector of the economy in which it operates – but plays a major part in determining the price of the company and so the likelihood of a takeover bid. Couple this with the detachment of many stockholders from any direct relationship with management, and it is not surprising that the prospect of making capital gains from takeover bids becomes a dominant consideration of the investor and his advisers.

In Germany, France and to a lesser extent in Japan, the finance for industrial companies is generally provided by a relatively small number of large investors and both the ownership of the rest of the stock, and the market in which such stock is traded, is much less important in the conduct of the company. These large investors may well be represented on the board of the enterprise by directors not dependent on public statements for their information on the policies and progress of the company and so 'locked in' to an entirely different extent to its activities.

The relative merits of these two different systems from the standpoint of stimulating management performance have been hotly debated. The arch takeover specialists like Hanson and Goldsmith vigorously promote the UK and US system's virtues while the opposite view is supported by comparing the current performances of the UK and US manufacturing industries with those of Germany and Japan.

The civil aircraft industry

It is not surprising that company reports do not go out of their way to give information on the nature and scale of their failures in product innovation. We get hints that a new drug development has been terminated, that a new model motor car is not achieving anywhere near its sales target and that a new piece of consumer electronics is a 'flop'. The exceptions, when they occur, usually relate to companies with a small product range – even a unique product – where the failures produce such a big impact on their financial resources that the details have to come out in the consequential reconstruction of their business.

The civil aircraft industry has produced more than its fair share of highly publicized failures of this type. In the transition from being a predominantly military aircraft business, in which the government carries the great majority of the risk, the companies involved had to pass through the danger zone of a single-product enterprise. Many failed at this stage and retreated from the business altogether. Some, having achieved success with their first product and even its successor, were brought down later before they achieved what we refer to as the 'critical mass' required to be established in this market.

Risk sharing with national governments has played a major role in the development of the industry. To begin with, in the 1950s, the development of civil aircraft and their engines was heavily subsidized in this way. Budgetary restraints, the claims of other sectors about the inequity of special treatment for aircraft companies and the manifest arrival of a big lucrative market for the successful caused support progressively to be withdrawn. This withdrawal, however, took an uneven form with some countries, such as the United States, retaining a 'de facto' support through massive military programmes and through government-funded research programmes. Others, in Europe, maintained partial launch aid for selected projects – the most prominent of these being the support of the Airbus Industrie as it developed, in the last twenty years, a range of products capable of competing with US companies.

Government support, in these circumstances, can be a fickle uncertain affair in which commercial decisions can be delayed and the aid eventually given less than that required. It can be given lavishly to such commercially-questionable products as a supersonic civil airliner and withheld, as it was in the UK, for the British Aircraft Corporation's competitive wide-bodied airliner – the BAC 3-11 – or for UK participation at the start of the Airbus project. It can be argued that companies that are dependent for a major part of their finance for innovation on government support experience an added risk as a result of these uncertainties.

To illustrate that 'betting the company' is not just a catch phrase but descriptive of real situations that arise in the pursuit of major product innovations, we have selected three well-publicized examples that have arisen in the aircraft industry. We start with the dramatic struggle for survival of the Boeing Commercial Airplane subsidiary of the giant Boeing Company.

The Boeing collapse

Boeing Commercial Airplane Company, which provides the majority of the sales of the Boeing Company alongside its military activities, is almost synonymous with the civil air transport industry. It currently has well over two-thirds of the international airline business – a near monopoly of the global market.

It was catapulted from a mediocre position in the 1950s by a combination of circumstances of which it took full advantage. The de Havilland Comet, having trailblazed the idea of jet transport, suffered a mortal blow in the tragic accident in which the fuselage burst due to metal fatigue. Unlike the UK defence specifications which demanded very specialized ultra-high altitude bombers, the US military had launched Boeing into the development of the B-47 where long range, even at the expense of take-off performance, was a dominant feature of the requirement. While the UK requirement led to the V-bomber production, the configuration required, typified by the Vulcan delta wing, was quite unsuitable for derivation into an economic civil airliner. A massive order for B-47 bombers, followed by an order for a jet-propelled tanker aircraft to flight-refuel fighters – designated later the Boeing 717 – gave Boeing the platform from which the classic Boeing 707 airliner was launched. In competition with the Douglas DC-8, the boom in civil jet transport began.

This made Boeing Commercial a very successful one-product company. The agonizing at how to follow this then began. The 707 was essentially a long-haul aeroplane and there were gaps lower down in range and size which had been created by the enthusiasm for jet transport and were in danger of being filled by competitors. In short order Boeing launched the '200 seater' Boeing 727 (entering service in 1964) and the '100 seater' Boeing 737 (entering service in 1968) to compete with the Hawker-Siddeley Trident, the Douglas DC-9 and the British Aircraft Corporation BAC 1-11. Finally, completing a comprehensive product range, Boeing, having lost to Lockheed in the funded competition for the big US military transport the C-5 (later named the Galaxy), went on to launch the jumbo Boeing 747 (entering service in 1970).

Throughout this amazing decade, Jack Steiner played a leading part as designer, project director and vice-president on most of the civil programmes. In 1974, in a paper to the Royal Aeronautical Society in London entitled 'Problems and Challenges – A Path to the Future', he gave a graphic account of the fight to survive that hit Boeing in the late 1960s (Steiner, 1974). It opens with the following statement:

> The market part of the story really starts in 1968, when the first signs of impending doom became evident. We at Boeing-Commercial had been flying along filled with the euphoria of being an expanding industry – the confidence which comes from having a good product line, good sales volume, an experienced management team and an excellent physical plant. Commercial orders were running at between two and three billion dollars a year. Our major product and plant decisions were behind us. We had invested close to a billion dollars in facilities over a four-year period and had assembly areas and machine capability unmatched in the world. Suddenly, in 1968, orders became harder to get, money tightened, the US market, which was two-thirds of the total, simply went away ... at the bottom we did not sell a single commercial airplane to a US trunk airline for a period of seventeen months.

Sales in the mid-1960s had been running at over 300 aircraft a year, so the collapse was quite disastrous.

There followed what must be one of the most courageous fights back to be successfully completed in the private sector. (There were, Steiner remarks in his paper, few votes in a government rescue for a company in Seattle!) It followed the classic formula of cutting costs and increasing sales and is fully described in Steiner's paper. A measure of its scale, in economic terms, is that employment in the Commercial Airplane side was reduced from over 80,000 in 1968 to 20,000 in 1971 while, simultaneously, management challenged every element of cost, every aspect of inventory and production control, and attacked every adverse sales feature of their product range (the Airbus and Douglas DC-10 were offering the wide body look).

The one thing that the paper does not reveal is the extent to which 'patient' money had to be secured to see Boeing through this transformation. Steiner goes this far, however:

> We have never revealed how close we got to the edge, or the peak number reached by our corporate debt. Our 1973 annual report did note that we paid off $474m of debt in that year. We obviously started paying it off long before that and on first quarter 1974 reported a continuation of pay off at about the same rate.

From this it is not unreasonable to conclude that at its peak an exceptional debt of the order of half of the annual sales of Boeing Commercial and close to its net worth prior to the collapse had arisen.

At Boeing's low point in morale, Steiner cites:

> a billboard on the road from Seattle to the Sea-Tac commercial airport. Paid for by two disgruntled employees it showed a light bulb on a wire and captioned in huge letters:
>
> WILL THE LAST PERSON LEAVING SEATTLE PLEASE TURN OUT THE LIGHTS

Not so – Boeing survived this massive debt to become the outstanding market leader in trunk route aircraft business that it is today.

The Rolls-Royce bankruptcy

Unlike the Boeing story, the events leading to the Rolls-Royce bankruptcy are exposed in clinical detail in an official report (Rolls-Royce, 1973) prepared for the UK Secretary of State for Trade and Industry by a lawyer and an accountant, appointed as inspectors under a section of the Companies Act. The report runs to some 400 pages and covers every aspect of the events leading up to the Board, in February 1971, ordering that dealings in Rolls-Royce shares on the London Stock Exchange should be suspended. Late the same day the trustee for the debenture holders appointed a receiver.

That Rolls-Royce was indeed 'betting the company' from the start is made plain in the report. By 1960, with substantial financial support from the Government, Rolls-Royce had built up a strong presence in the civil aviation market with the Dart turbo-prop engine, the Avon jet engine in the French Caravelle, the Conway by-pass engine and was launching, now with significant investment of company funds, the Spey fan engine for the BAC 1-11. As early as 1962, however, Lord Kindersley, then Chairman of Rolls-Royce, found it necessary to warn the then Ministry of Aviation in the following terms:

> The financial position in which the Aero-engine Division of Rolls-Royce finds

itself is basically caused by the amount of its own money which it has sunk in civil engine development and the launching costs of civil engines...

It is particularly disappointing to us that, while Rolls-Royce has been risking its own money in the development of civil engines, it has received a steadily declining proportion of the Government money being spent on military engine development.

We have been told on more than one occasion that special consideration would be given if the future of our business was in jeopardy because of lack of finance. In my opinion this position now exists.

The concern expressed in this letter on the lack of military orders refers to the accumulation by Bristol-Siddeley (a jointly owned Bristol Aeroplane and Hawker-Siddeley aero-engine company) of most of the military engine orders placed in the early 1960s by the British Government. However, worse from Rolls-Royce's point of view was to come. A derivative of the Bristol-Siddeley Olympus engine used in the Vulcan bomber was chosen to be jointly developed with the French engine company SNECMA for the Anglo-French Concorde airliner. When it, further, became clear that these two companies were discussing a joint venture with Pratt & Whitney (Rolls-Royce competitor in the USA) to supply a large fan engine for the Airbus, Rolls-Royce opened discussion with Bristol-Siddeley's owners to buy the company. By the end of 1966, Rolls-Royce had acquired Bristol-Siddeley for £27m in cash and £37m in shares. Whatever the benefits strategically of this acquisition, the impact of it on the already weak Rolls-Royce financial position was to make it weaker.

Lord Kindersley's last paragraph referred to stated Government policy in the early 1960s that it would back the aircraft industry in playing a prominent part in the development of the worldwide civil aviation market. However, by the middle of the 1960s, Government-commissioned studies were questioning such a policy and there was a grudging reluctance to give support (with the extraordinary exception of the Concorde) to civil aircraft related programmes. This reluctance was underlined when the then Minister of Aviation chose the occasion of the Society of British Aircraft Constructors annual dinner in 1967 to remark that the industry had obtained sums of money which 'made the Great Train Robbers (reference to a daring theft of currency in the news at the time) look like schoolboys pinching pennies from a blind man's tin.'

It was in this financial scenario that Rolls-Royce embarked on the development of a totally new product, the RB 211. Much has been made afterwards of the technical problems involved and, with hindsight, Rolls-Royce suffered from a leadership vacuum when its brilliant Technical Director, Adrian Lombard, died suddenly at a crucial time in 1967. Nevertheless – and despite the impossible climate in the year preceding the bankruptcy – on the day the company was declared insolvent the RB 211 produced close to its contractual performance and, as history shows, went on to demonstrate the lowest fuel consumption and best performance retention in service of all the big fan engines. The root cause of failure was not technological but was due to inadequate financial resources. The report referred to above makes this very clear.

First, the RB 178 demonstrator engine programme – the accepted 'skunk works' approach to flushing out the technical problems of a major technical advance ahead of committing to full development – was terminated 'to save money' with a number of problems left unresolved. Then launch aid from the Government was made conditional on securing a firm application for the engine. In such circumstances the buyer, Lockheed, had the whip hand and extracted price and delivery dates which were extremely difficult to meet. Finally, the amount of aid Rolls-Royce sought was plainly stated to be that required over and above what Rolls-Royce could afford. When it came, after prolonged negotiation, it was less than Rolls-Royce forecast would be required (and, therefore, in the nature of innovation, much less than what *would* be required).

Thus the technical development started at a disadvantage, was conducted under the twin threats of dire consequences if Lockheed's delivery dates and prices could not be met and manifestly inadequate funds to cover the development cost and the initial losses on the early production engine deliveries.

By 1968, having raised further equity of £21.7m, the company balance sheet showed a net worth of just over £100m with £57.2m in debentures and loan stock and bank indebtedness of £32.4m. The company's estimate of gross launch cost was then £115.10m with a fixed sum of £47.0m contributed by Government. Thus some 60 per cent of the net worth of the company was planned to be at risk on the RB 211 project alone. With any reasonable factor for the unknowns of innovation, this meant the entire net worth of Rolls-Royce was being placed at risk – truly 'betting the company'.

Rolls-Royce clearly did not have anything like the 'critical mass' to embark on the RB 211 on its own and misled itself into believing that sufficient launch aid would be forthcoming to overcome this. On the premise that if the market was there (as it has subsequently turned out to be with some 1500 RB 211 type engines sold) finance would be forthcoming, the company went ahead. As the Government inspectors themselves said in their report:

> The protection of the public purse is, as far as it goes, a commendable objective. What may be more open to question is the wisdom of making an investment of public funds sufficiently substantial to encourage a company to commit itself to a project, the estimated costs of which by its very nature are liable to be exceeded by a wide margin, while at the same time so limiting the investment that if that margin proves to be such as to place the company itself

in jeopardy, both the company and the public's investment may be wiped out.

The Westland rescue

In the summer of 1985, it became public knowledge that Westland plc, a group of companies with the majority of its business in helicopter manufacture, was in financial trouble and that an injection of new capital was required if the company was to continue trading.

The story starts in the 1960s when, under pressure from the Government of the day to rationalize the diversity of defence-related enterprises in the UK, Westland acquired the helicopter activities of Fairey Aviation, Saunders-Roe and Bristol Aeroplane Company and became the sole source of helicopter design and manufacture in the country. Soon after came the so-called Anglo-French package deal, in which the two Governments agreed to finance the development and production of three new types of helicopter – the Gazelle, the Puma and the Lynx. The two 'national champions' of the two countries were put together to execute this programme. Production would be shared between them in accordance with an agreed formula (which included export as well as national forecasted sales) with design leadership for the Lynx being assigned to Westland and for the Gazelle and Puma to the Helicopter Division of what is now called Aérospatiale but whose full title is Société Nationale Industrielle Aérospatiale, the French state-owned aerospace company.

By 1975, this programme had reached the point when both companies faced the problem of what new projects should be started to replace their current product range for the 1980s and beyond. Vigorous attempts, first à deux and then with the other two Western European helicopter manufacturers (Agusta of Italy and MBB of Germany) to encourage their respective Governments to continue to support the industry by, at least, financing collaborative pre-competitive research and demonstrator programmes, were made. Meetings with officials were established and rotated in the capitals of the four countries but failed to come up with anything that met the imperatives of action facing the four companies. While they continued to be held, the companies had no alternative but to 'break ranks' and seek to solve their particular problems in other ways. Aérospatiale, with strong Government support, launched the Dauphin; Agusta the A109 and A129; MBB formed an alliance with Kawasaki of Japan to launch the BK 117; and in 1976 Westland launched its Westland 30 series.

The future strategy of which the Westland 30 was a part also included the EH 101. This latter type emerged from the quadripartite studies as a replacement for the Sea King type of helicopter used by many navies as the carrier of submarine detecting and attacking systems. While German naval requirements were minimal and France accorded them little weight in their defence priorities, both Italy and the UK were likely customers for an advanced helicopter-based system to match the increasingly difficult task of locating modern submarines. Westland and Agusta set about the task of securing their respective Governments' interest in a Sea King replacement which, it was hoped, would find favour as a jointly financed European defence project.

Budgetary pressures in both countries, coupled with the relatively small numbers each navy required, soon revealed that a fully defence-financed project was very unlikely to get support. A search then started for other applications and other sources of finance which could be put together and a more economically viable case developed. A transport derivative with a rear-loading ramp, a thirty-passenger commuter civilian variant and an offshore drilling rig service vehicle were all designed and their sales prospects supported with both in-house and market consultants' studies. Lobbying, speculative company investment and the pursuit of government funds for the civil variants extending over a number of years culminated in the launch of full development through a joint Westland–Agusta company named EH Industries Limited in 1983. Funding was secured from six sources in roughly equal amounts – the British Ministry of Defence, launch aid for the civil version from the British Department of Trade and Industry and from Westland's own resources, with the three counterparts of these provided on the Italian side.

This gave Westland a product for the 1990s and well into the next century but left a bleak prospect for the 1980s and, more importantly, no prospect of the cash-positive consequences of the 1980s offsetting the inevitable cash-negative impact of the introduction of the EH 101 in the early 1990s.

There could be no question that Westland could either finance or develop in time a new competitive helicopter for the 1980s. On both grounds, therefore, a programme of low cost step-by-step development of derivatives, starting with the Lynx rotor system on a new fuselage, was decided upon. It started with a brand new, design-to-cost fuselage capable of carrying seventeen passengers plus a generous luggage compartment, or thirteen fully-armed and equipped troops. This was initially powered (or more correctly underpowered) by the production Lynx engine transmission and rotor system. With some uprating of the Rolls-Royce Gem engine it was offered as a low first cost per seat, reliable short range transport with a much better standard of passenger comfort (the new fuselage) than the noisy and cramped cabins of the competition. It went into service with British Airways Helicopters carrying crews to the gas production rigs in the North Sea, in Pan American's service from Kennedy Airport to the Manhattan heliport and with a similar operation in Los Angeles. It was also chosen by Mrs Indira Gandhi's administration in India to be bought for servicing India's offshore oil rigs.

With this evidence of company initiative – and a full civil airworthiness certificate for the Westland 30 series 100 (as the first model was called) – secure, Westland in 1981 approached the Government for a contribution to the financing of the development of the Westland 30 – 200 and 300 series. The case was heavily documented with company financial information and with an encouraging independent market survey by a prominent New York market research organization, who put its 'most likely' forecast of sales up to 1990 at over 300.

The company set out the strategic context in which this application was made in the following terms for support for Westland Helicopters (WHL):

> It is the policy of the Westland Group that WHL should exploit the civil helicopter market to secure future growth, improved financial performance and better employment prospects. This is not only because of the civil market growth prospects which will strengthen the base of the company, but because the disciplines of designing and operating in the commercial marketplace will make a substantial contribution to success in future military transport programmes. In the long term this strategy must embrace both Westland 30 and EH 101.
>
> Westland 30 series 100 has been developed at the company's expense as a derivative of the Lynx and is now in production but full exploitation of the market potential can only be achieved by the development of series 200 and 300. Series 200 will have improved capability at altitude and in high temperatures while series 300 will have an extended range.
>
> The EH 101 is a new aircraft and the cost of developing it will therefore exceed that of the Westland 30, even with international collaboration. Government participation in its development will be essential from the start. In preparing the financial forecasts submitted as part of this proposal it has been assumed that any part of the development costs of EH 101 which the Group agrees to bear will be contained within the Research and Development appropriations shown in Appendices 2 and 3. Since the success of both products is dependent on their availability for delivery in the timeframes determined by the respective market surveys, both projects must proceed concurrently.
>
> The chosen course of action is therefore to pursue further development of Westland 30 for the civil market of the 1980s and beyond, with EH 101 following on at the end of the 1980s and continuing in production into the next century.
>
> The financial implications of that course of action are considerable. In addition to the development and launch costs of new helicopters for the civil market, success depends heavily on the ability to deliver aircraft and spares on short lead times. This will require the manufacture of aircraft in advance of orders and the maintenance of a comprehensive stock of spares for prompt product support – both without the pattern of deposits and progress payments common in the military market. In addition, marketing costs will be considerable and a network of service facilities will be required at strategic points around the world to give operational security to the customer.

The changed financial requirements arising from this new business environment need to be understood before considering the case for supporting development and launching costs.

The cumulative cash flow profile of the project was presented. It took the form, typical of aircraft projects, illustrated in Figure 2.2. Even with the launch aid requested, it was made plain that additional equity capital would be required of some £50m to give reasonable headroom and keep gearing down to some 33 per cent.

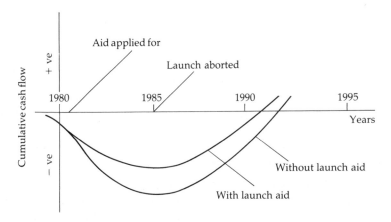

Figure 2.2 *The form of projected cash flow – Westland 30 Project*

The 1981 Annual Report of the Company (Westland Aircraft, 1981) showed shareholders funds of some £115m with long-term loans of £7m and under £4m short-term borrowings. By 1984 long-term loans had reached £35m with short-term loans and overdrafts at £24m. The Westland 30 venture had gone seriously wrong for a combination of reasons:

1 The market for civil helicopters of all types, which had boomed as the rush to develop offshore oil resources following the oil crisis in the early 1970s, dried up and all manufacturers were in serious trouble. Agusta, Westland's partner in the EH 101, had to be rescued by the Italian Government with a large inventory of unsold A109 helicopters. Bell, Sikorsky and Aérospatiale all reported very low sales and financial difficulties but were cushioned by being part of larger companies and, in the case of Aérospatiale, by being a state enterprise.
2 The niche of this market, the short-range passenger market servicing airports, to which the Westland 30 series 100 was in part directed, failed to develop. The New York service continues to operate but the Los Angeles operation folded when the city authorities failed to provide the projected downtown terminal. Chicago's O'Hare service failed to start.

In the UK, the Heathrow–Gatwick shuttle was stopped on the grounds of noise and the alleged adequacy of the new M25 highway linking the two airports.
3 The cost of supplying and supporting these pioneer ventures, which would have been justified if they had burgeoned, turned out to be a serious source of loss when they failed.
4 The intent by the Indian Oil and Natural Gas Corporation to buy twenty-one Westland 30 100 was not turned into a firm contract after Mrs Gandhi's assassination and, because of the short delivery dates required in the competition, Westland was left with a large inventory of unsold Westland 30 100 helicopters (subsequently ordered after the Westland 'rescue').

To add to this chapter of cash-absorbing setbacks, a military contract got into difficulties – not with the helicopter element but with the supply and functioning of some complex subcontracted electronic systems. This left a further increment in inventory to be financed and a provision to be made for the increased costs involved.

The Chairman's statement for 1984 contained the following in respect of the civil helicopter market (Westland, 1984):

> It may appear from this year's results that it would have been better if we had kept out of the civil market. But in fact the defence markets open to us are simply not large enough to back a viable production operation, and are certainly much too small to justify a new research and development project. Without entry into the civil market we could never have justified the EH 101 project, nor would the British and Italian Governments have contemplated it. Naval sales alone of EH 101 are not expected to exceed 300; and that would not justify the launch of the development of a new helicopter.
>
> But there is another reason for Westland being in the civil market: the lessons that come from fierce competition, and from the far more intensive use of the helicopter by civil operators than in the defence forces except in fighting conditions. Westland has never been just a subcontractor to a licensor; we have long had a strong engineering department enabling us to innovate, improve, adapt for customer requirements and where desirable pool our skills with others. No one else in the United Kingdom has that capability.
>
> We knew, and said when we started into the civil market that we had to face a strain on our Profit and Loss Account from the considerable addition to the expenditure on private venture research and development for some years. We sought to provide for this in the only way modern accounting practice allows by setting up a development reserve, from which we are drawing this year. We did not appreciate to the full the costs of developing the market for a new helicopter in a new role; and we had not foreseen the depth of the helicopter recession. We had foreseen, and provided adequately for, the learning costs of the Westland 30. And we had foreseen the need to carry a high inventory of components so as to give customers their required

short delivery times. With all these lessons and considerations we are now well poised for a full recovery of the market, albeit in the immediate future to a level lower than we expected three years ago. And, I should add, that we shall be all the better able to tackle the launching of the EH 101.

In April 1985, a hostile takeover bid was launched by a consortium, Bristow Rotorcraft plc, formed specifically for this purpose. It was led by Mr Alan Bristow, one of the world's most successful helicopter operators. Much was made in the offer of the company's difficulties but the attack was focused on the Westland 30. Its civil prospects were rubbished and doubts cast on its prospects for a UK military requirement which had been part of the Westland strategy. The offer did, however, promise an injection of £60m in cash.

In a letter to shareholders (Westland, 1985a), the Chairman of Westland reluctantly recommended this offer in the following terms:

> While the Board still feels the long-term future of Westland would be better served by an association with a substantial international business it is clear this is not currently available.
>
> Under the BR [Bristow Rotorcraft] proposal . . . would result in a £60m cash injection. Your Board considers it is not currently possible for Westland to raise such a sum on these terms from existing shareholders.
>
> In the absence of a better deal, your Board and its financial advisers . . . recommend you accept the BR offer.

This went out on 13 June 1985. On 20 June, Bristow Rotorcraft withdrew their bid. The situation so created could not have been worse for Westland, and drastic action had to be taken to restore credibility. A new chairman took over and, in the words he used in his first annual report (Westland, 1985b), described the action then taken in the following terms:

> My predecessor . . . stated that in the Board's view the Company required on association with a substantial international business.
>
> In discussions with both the Ministry of Defence and the Department of Trade and Industry it was made clear to me that, although Westland was considered a Company of strategic importance, public funds would not be made available to assist with a financial reconstruction and that a private sector solution would have to be found.
>
> . . . the Board unanimously recommended that the proposals based on association with United Technologies and Fiat provided the best prospects for the medium and long-term future of the Company.

The further development of the Westland 30 – the series 300 which was to have been the competitive model – was terminated and £80m written off as exceptional provisions against profit. The reconstruction was successfully carried out, though after a quite unnecessary amount of public drama in which two senior Government Ministers resigned. In the year after the reconstruction, 1986, Westland turned in a profit before tax of £26.4m on sales of £344.4m – a return on net operating assets of 28.2 per cent (Westland, 1986).

The concept of critical mass

We began this chapter by considering the very real risks of innovation and have ended it with the consequences for companies who get it wrong, for whatever reason.

What is striking about the examples we have chosen is the benefits to be gained by what might be called an 'internal solution' to the problems involved. In both the Rolls-Royce and Westland affairs much diversion of effort arose from having to parade the problems in the external financial marketplace with its legalistic procedures and political overtones. Boeing managed to avoid this by securing 'patient money' to see it through its years of reconstruction.

This 'patient money', accompanied by drastic internal management action, can most readily be secured if the finance required for a specific innovation is only a fraction of the net worth of the enterprise. Both Rolls-Royce and Westland effectively were betting their net worth, and a loss in this situation immediately puts the matter into the insolvency category. The phrase 'critical mass' is now entering the vocabulary of industry as the size of enterprise required for this condition to be averted. We will examine this further in Chapter 3.

References

BBC (1986), *The Pace of Change*, BBC Radio 4 Analysis Series, Talks and Documentaries Department, BBC London.
Rolls-Royce Ltd (1973), *Investigation under Companies Act 1948*, HMSO.
Steiner J. E. (1974), 'Problems and Challenges – A Path to the Future', *J. Royal Aeronautical Society*, July.
Westland Aircraft Ltd (1981), Annual Report.
Westland plc (1984), Annual Report.
Westland plc (1985a), Chairman's letter to shareholders, 13 June.
Westland plc (1985b), Annual Report.
Westland plc (1986), Annual Report.

3 The pursuit of critical mass

- The scale of the problem
- The scale of the answer
- The automobile industry
- The automobile component industry
- White goods – Electrolux and Whirlpool
- Electrical power engineering – ABB and GEC–Alsthom
- The pharmaceutical industry
- Telecommunications
- Final remarks

The scale of the problem

We have seen in the previous chapters that innovation is both expensive and risky. When the costs reach the order of magnitude of the net worth of a company and, for whatever reason, things go wrong, then the very survival of the enterprise is in jeopardy.

The direct consequence of this is that prudent management, finding themselves between the Scylla of the need to innovate to survive and the Charybdis of putting the company in jeopardy if they do, are driven to pursue the so-called critical mass solution. This solution seeks either to reduce the amount at risk or to raise the financial strength of the enterprise to the point where, in their judgement, survival of the company is not being jeopardized. It involves mergers, takeovers and joint ventures in which the complexities of compatibility of management style, market coverage and technological synergy play a decisive part; matters on which we try to shed some light in Chapter 5. Here we open the subject by attempting to put a rough scale to the problem and then examine a range of solutions which have come into the news in recent years.

We start by assuming that the amount that is prudent for a company to invest in launching an innovative product is proportional to the net worth, which in turn is in proportion to its annual sales. Since we have a 'conceptual' formula for total launch investment of $L = P_c V_c$, itself measured in units of annual sales, we immediately see that prudence first demands a multiplicity of products at various stages from launch to obsolescence. Without this, 'betting the company' is unavoidable.

The second consequence, as we have seen in the examples given in Chapter 1, is that very large enterprises are required to compete in such markets as civil aircraft, mainframe computers, etc. The price–volume criterion for 'criticalness' is, of course, no more than a measure of the entry barriers to the market concerned – a minimum stake is required before the company concerned can enter the gambling game, let alone win.

With all its inaccuracies, the critical mass diagram shown in Figure 3.1 does enable an overview of the critical mass problem of innovation to be obtained. It has been produced by taking, arbitrarily, $3\,P_cV_c$ as the minimum annual sales required before it is 'prudent' to launch an innovative product with a characteristic selling price P_c and a characteristic sales volume V_c. This is its crude arithmetical origin but, having created it, it is intended to be used solely as an indication of trends.

Superimposed on this diagram are three categories of company, identified as follows:

- *Macroindustry* With annual sales of billions and an innovative risk capability of hundreds of millions (probably manufacturing in many countries of the world and employing hundreds of thousands of people).
- *Industry* With annual sales of hundreds of millions and an innovative risk capacity of tens of millions (probably manufacturing in a single country but exporting worldwide and employing tens of thousands of people).
- *Microindustry* With annual sales up to tens of millions and innovative risk capacity of a million or less (nation-based, employing up to hundreds of people and extending downwards to the embryonic innovative enterprise in the venture capital field).

These are obviously only broad benchmark categories. The terminology may need some defence.

Industry, as defined above, is the generally recognized category. In general parlance it is the image that is immediately called up when the subject of 'industry' is referred to. Below this, there has grown up a recognition of the small company with its own managerial culture and place in the hierarchy. This we have given the name *microindustry*.

However, above the level of industry, general parlance degenerates into a variety of terms implying a mixture of incomprehension, supranational power and commercial dominance. 'Mega' has been rejected as a prefix to describe this industrial sector because it projects bigness as a direct consequence of the market and not as an end in itself. The considerations which lead to it are quite different from those that lead to conglom-

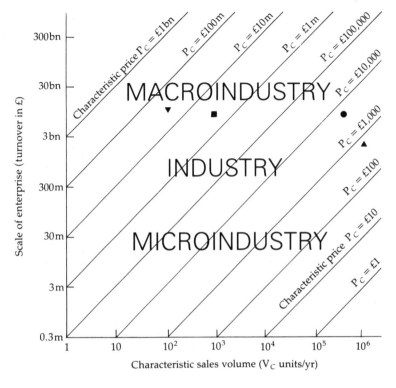

Figure 3.1 *Critical mass diagram – minimum scale of enterprise to launch an innovative product in a market with characteristic sales volume (V_c units/yr) and characteristic selling price (P_c)*

erates of the Hanson Trust type, large groupings of disparate companies whose connecting links are almost exclusively financial. These are often big and may well be internationally spread but they can be assembled and dismantled ('unbundling', as it is called) at a pace determined by the procedures of the sale or purchase of shares. In contrast, the enterprises required to pursue competitive innovation on the grand scale involve integrating some or all of the elements of the innovative chain from the technology base through into worldwide marketing. Creating these to achieve a new product may take a decade or more and reaping the benefit a great deal longer. Etymologically, μακρο rather than μεγα conveys this time element.

Macroindustry is also attractive as a name because, as we seek to show later in this book, it poses a new challenge to political economy in general and to that area covered by macroeconomics in particular. It is much more than industry written big.

The importance and position of macroindustry in the innovation scene is by no means confined to the companies at this level. Innovation is not of course restricted to the final assemblers of a product like an airliner, an automobile or a computer. Into each go a whole range of manufactured products, such as the proprietary aero-engines, navigation and auto-pilots and the specialist materials which make up a modern airliner. The input of all these plays an essential part in the competitiveness and success of the final product. The markets for these have their own individual price–volume characteristics. These, in turn, then decide the critical mass they require to be competitive in their field.

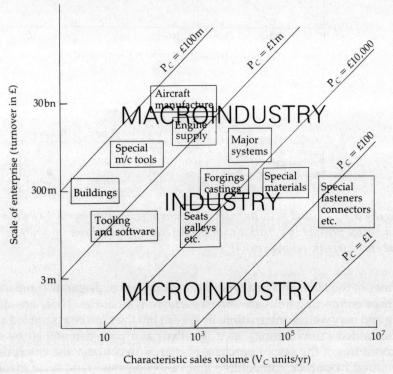

Figure 3.2 *Aircraft manufacture – critical mass of a range of innovative suppliers*

Figure 3.2 illustrates this by positioning a typical range of supplier companies to an airliner manufacturer on the critical mass diagram. The position of the vital specialist machine tool and testing equipment manufacturer on the one hand and the aero-engine and proprietary system supplier on the other is clearly shown.

While the critical mass for many of these innovative activities may well lie comfortably within the industry category of company size, it must be remembered that they will depend for their survival on the macroindustrial enterprises they supply. A national innovative industry without access, for whatever reason, to supplying the macroindustrial companies who command key industrial sectors will wither and die. The possibility of this arising and the measures required to deal with it will emerge as one of the challenges governments face in the global innovation scene.

The scale of the answer

Table 3.1 contains a selection, graded by annual sales, of the top companies of the world in competitive innovative manufacturing and grouped into the so-called triad of geographical origin – USA, Europe and Japan. The selection was made from the *Times 1000*, (Times, 1989–90), an annual publication and, of course, annually made out of date by the restructuring that is going on even at this level. The assumptions used in the selection were:

1 Annual sales exceeding £4bn (exchange rate around $1.6 to £1), 1989 economic conditions.
2 Excluded are the oil companies like Exxon, Shell, BP, Elf Aquitaine etc., retail chains such as Sears Roebuck and the massive state holding companies like IRI (£40bn).
3 More arguable, in our context, is the exclusion of companies such as BAT Industries (£23bn) and Veba (£22bn) on the grounds that their innovative activities are only a small and fragmented part of their 'bundle' of companies. British Telecom (£17bn) has also been left out as an, as yet, undeveloped presence in the manufacturing scene.
4 In the Japanese section, the list omits the powerful trading alliances – the sogo shosha – but includes their manufacturing subsidiaries. Mitsubishi, Mitsui, C. Itoh, Marubeni and Sumitomo Corporations are all listed as having sales in excess of £65bn. For example, Mitsubishi Corporation (£70bn) integrates the marketing, financing and other activities of the three enterprises we have listed, namely Mitsubishi Heavy Industries, Mitsubishi Electric and Mitsubishi Motors.

These seventy-five companies, covering macroinnovative capabilities in the chemical, pharmaceutical, electronic, data processing, power

Table 3.1 *The macroindustrial innovative companies of the world*

USA (28 companies)

General Motors	(70)	McDonnell Douglas	(10)	General Dynamics	(6)
Ford Motors	(59)	Rockwell	(8)	ITT	(6)
IBM	(38)	Allied Signal	(8)	Motorola	(5)
General Electric	(25)	Westinghouse	(7)	Raytheon	(5)
A T & T	(22)	DEC	(7)	Honeywell	(5)
Du Pont	(21)	Goodyear	(7)	TRW	(4)
Chrysler	(20)	Lockheed	(7)	Emerson	(4)
United Technologies	(11)	Caterpillar	(7)	Texas Instruments	(4)
Eastman Kodak	(11)	Unisys	(6)		
Boeing	(11)	Hewlett-Packard	(6)		

Europe (29 companies plus 3 US subsidiaries consolidated in Ford, General Motors and IBM figures)

Daimler-Benz	(24)	Peugeot	(11)	Rhône-Poulenc	(6)
Siemens	(19)	ABB	(11)	British Aerospace	(6)
Volkswagen	(19)	Volvo	(9)	GEC	(6)
Fiat	(17)	Bosch	(9)	Michelin	(5)
Philips	(16)	BMW	(8)	BTR	(5)
BASF	(14)	Saint-Gobain	(8)	MAN	(5)
Renault	(14)	Thomson	(7)	Saab-Scania	(4)
Hoechst	(13)	Electrolux	(7)		
Bayer	(13)	Ciba-Geigy	(7)	* Ford	(6)
ICI	(12)	Mannesman	(7)	* Opel	(6)
CGE	(12)	Montedison	(6)	* IBM	(4)

Japan (18 companies)

Toyota Motor	(30)	NEC	(11)	Sony	(6)
Matsushita	(18)	Mitsubishi Electrical	(10)	Nippondenso	(5)
Nissan	(16)	Fujitsu	(9)	Isuzu Motors	(5)
Hitachi	(15)	Mitsubishi Motors	(9)	Sharp	(4)
Toshiba	(13)	Mazda	(8)	Sanyo	(4)
Honda	(12)	Mitsubishi Heavy Industries	(8)	Suzuki Motors	(4)

Figures in brackets are annual sales in £bn
Exchange rates ($1.6 = £1)
Source: *Times 1000* (1989–90)

generation and aircraft and vehicle manufacturing sectors have an aggregate annual sales of some £1000bn. They already have a world market share in their sectors approaching percentages which, on a national scale, would provoke anti-trust concern; and, as we shall now see, this is only a snapshot of an ever-increasing process of aggregation by merger, takeover and joint venture. Let us examine the recent record of this activity in pursuit of critical mass as derived from public announcements. It comes in the form of a 'scrapbook' picture, flavoured with investigative journalistic comments that are inevitable when the source of information is the financial press, which all in all seems more revealing than the circumscribed and cautious accounts to be found in company reports.

The automobile industry

Of all the manufacturing sectors, the automobile industry has for many years been conducted in large industrial units. However, with the exception of Ford and General Motors, who have had significant plants outside their home countries for over half a century, these have been essentially national with an export dimension.

The present scene, however, of intense internationalization, is a relatively recent development. This has brought with it a burst of competitive innovation, led by the Japanese, and competitive innovation, as we have shown, means high investment and ever larger markets to recover it.

Not that the manufacturers of the popular brands of motor vehicles are not already very large. In the table of seventy-five companies (Table 3.1), no less than eighteen are immediately identifiable as motor vehicle manufacturers and the number goes up to twenty-three if those whose major outlets are vehicle components and vehicle tyres are included. However, what is not so readily appreciated is that, massive though they may be, all are vigorously searching for critical mass through joint ventures and alliances. In a rapidly-changing situation, the already out-of-date picture of alliances in the automotive industry shown in Figure 3.3 still gives an idea of the complex situation in this respect in the industry (Devlin and Bleakley, 1988).

An example of a recent development in this area was announced in 1990 between Renault and Volvo in an attempt to ensure that each partner maintains his integrity and a large measure of independence, while aiming to 'gain considerable synergies and increased efficiency' by being 'well above the critical mass needed to compete in the world market'. Renault had held a minority stake in the Volvo car company for several years before the 1990 formal alliance was announced, and the two partners had flirted with the idea of a closer relationship for some time, culminating in an agreement that specified reciprocal share holdings, particularly in the trucks and buses business: with 45 per cent of Volvo trucks and buses held by Renault and 45 per cent of Renault's trucks and buses held by Volvo (significantly, there are reciprocal stake holdings in Volvo cars and Renault cars as well, but the level of these holdings is appreciably lower, presumably because the car business is much more profitable for both Volvo and Renault than trucks and buses), the partners clearly signal to the motor industry their determination to protect each

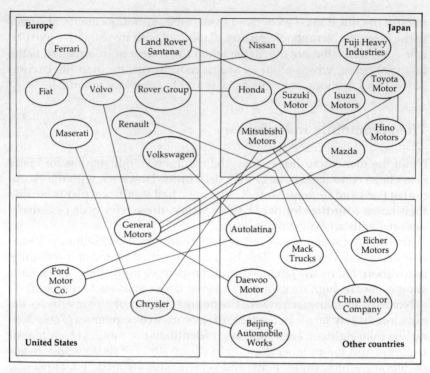

Figure 3.3 *Automobile industry – joint ventures and alliances*
Source: Devlin and Bleakley (1988)

other from external threats and to move towards close collaboration through an elaborate structure of coordinating committees. Foremost on the agenda is their declared intention to cooperate in the fields of research, technology, product development and production.

In outlining the rationale for the Volvo–Renault alliance, Mr Raymond Levy of Renault indicated several key issues looming on the horizon for the motor car industry:

- The danger of a downturn in sales, both in cars and trucks.
- Questions about 'the very acceptability of the automobile' as public concern grows regarding pollution and noise (with inevitable pressures to introduce demanding emission standards) and traffic congestion, particularly in urban areas, all of which are bound to have serious implications for engine and body design.
- The high cost of R&D and launch of new models (both in capital costs and market development).
- The need to complement each other's strengths in technology, design and product range, so as to be able to stand up to the competition and particularly to the Japanese threat.

Ford, in recent years, has been in the news for:

- A joint venture with Fiat (IVECO–Ford trucks).
- A joint venture with Nissan ('people carrier' vehicle development).
- A joint venture with Mazda (in which Ford has a 25 per cent stake) (Escort replacement development).
- A joint venture with VW in South America (Autolatina).
- The acquisition of Jaguar.

General Motors, with its recent merger with SAAB and its joint venture with Isuzu to make trucks in the UK, is active in all sorts of alliances, despite being already a £70bn sales company.

The need for further restructuring into larger units of the European automobile industry is accepted and has started. Faced with the 'Japanization' of the UK automobile industry through the establishment of the Rover–Honda joint venture, and the Nissan and now Toyota major investments, Europe's fragmentation into national fiefdoms – Fiat in Italy, Mercdes–Benz in Germany and Renault and Peugeot in France – is breaking up as the 1992 open market makes its mark. The Volvo–Renault potential of a £23bn enterprise, second to none in size in Europe and only exceeded in the global market by GM, Ford and Toyota, clearly points to the way the European car industry is heading.

The automobile component industry

The pressure for critical mass has inevitably extended into the car component sector. It comes on them from two directions. First, these manufacturers are under pressure to become responsible for whole systems (transmission, brakes, engine management, environment control, etc.) and to carry the burden of competitive innovation in their particular field. The second comes from the geographical spread of assembly plants round the world, which suppliers are expected to follow. The recent moves of Bosch, Nippondenso and other Japanese suppliers into the UK are clear examples of this trend.

As a result, this tier of manufacture is now rapidly moving into the macroindustrial category with TRW, Bosch and Nippondenso already there and Valeo (a European company recently created by merging a lot of small suppliers in Europe), Lucas and others not far behind.

This is also true of the tyre industry. The top ten tyre makers, with their sales in the tyre market assessed separately from their total corporate sales in 1988, are listed in Table 3.2.

At that time Bridgestone of Japan had just acquired the United States manufacturer, Firestone, Sumitomo had acquired the tyre manufacturing side of the UK Dunlop plants and Michelin were said to be seeking to

Table 3.2 *The top ten tyre makers (1988)*

Company	Sales ($bn)
Goodyear	8.0
Michelin	8.0
Bridgestone/Firestone	6.5
Continental/General Tire	3.8
Sumitomo/Dunlop	3.0
Uniroyal/BF Goodrich	2.0
Yokohama	1.7
Toyo	0.75
Cooper	0.75

acquire the aircraft tyre business of B. F. Goodrich, which had recently merged with Uniroyal.

Since then, and a measure of the rate of aggregation of tyre manufacture, Michelin paid $1.5bn to take over the entire Uniroyal–B. F. Goodrich company to put it ahead of Goodyear and making it a major supplier of original equipment to General Motors. All are agreed that the pressure of competitive innovation – not only in product design but in manufacturing technology – is the force driving them to bigger and bigger companies with a global spread of activities in line with those of their customers, the automobile manufacturers.

White goods – Electrolux and Whirlpool

Like the motor car and the television set, the market for technology applied to the work of running a home is now firmly established – and poised to grow. It is, therefore, not surprising that it, too, provides examples of the pursuit of critical mass.

The Swedish company, Electrolux, started its intensive pursuit of global critical mass in 1984 and a spate of major acquisitions followed. First was Zanussi of Italy. This was followed in March 1986 with it paying some $750m to win control of White Consolidated in the United States. White was then Number 3 in the US market for domestic appliances after General Electric and Whirlpool.

This was followed in November 1986 with the acquisition in the United States of Beaird-Poulon, the garden and forestry products division of Emerson Electric to add to its presence in this market through Flymo and Husqvarna. In April 1987, it acquired in the UK the domestic equipment subsidiaries of Thorn-EMI, retailing products with such brand names as

Tricity, Bendix and Parkinson-Cowan, making it the leader over GEC in this market in the UK. In September 1987, Electrolux acquired a minority stake in two Spanish state-owned domestic appliance manufacturers with an option to acquire control later. In November 1988, Electrolux acquired the garden equipment side of Roper, one of the US leading producers in this field, for $255m. As a result of this, Electrolux outdoor products division had annual sales in excess of $1bn.

Electrolux seems now to be engaged in 'digesting' these acquisitions. Mr Anders Schap, Electrolux's Chief Executive, in an interview in 1989, accepted that he faces some major management tasks, not least changing from a Swedish to an international style of management, but is adamant that he is building an integrated organization and not a conglomerate.

Whirlpool, the second largest supplier of white goods in the USA after GE, and Philips, the second largest supplier in Europe, have countered by combining their activities in this sector. Whirlpool has the controlling interest in what is now called Whirlpool International and reported, on its first year of operation, 'a satisfactory performance in this highly competitive market'.

Outside Japan, these two companies plus GE therefore dominate the world market in domestic appliances. The Japanese market is, as in other fields, highly protected but Electrolux, who have had a subsidiary in Japan since 1975, have recently established an arrangement with the Japanese company Sharp to try and gain more access through Sharp's distribution network.

Electrical power engineering – ABB and GEC–Alsthom

In August 1987, the proposed merger between Brown Boveri of Switzerland and Asea of Sweden was announced. It was described as a 'bolt from the blue' and as 'the most dramatic example so far of the new fashion for international mergers in Europe's Balkanized marketplace for electrical power generation.'

Both companies have a long history. Asea was founded in 1883 and developed into a group with some 350 companies employing 71,000 people in 100 countries. It had significant shareholdings in Electrolux, ESAB, a world leader in welding equipment and also had wholly-owned subsidiaries making all-terrain vehicles – Hagglund and Soner and hydroelectric equipment – Skandiaviska Elverk.

Brown-Boveri, Switzerland's third largest corporation after Nestlé and Ciba-Geigy, is one of the world's largest electrical engineering groups. Founded in 1891 by an Englishman, Charles Brown, and a Bavarian,

Walter Boveri, to build a power station, it employs some 100,000 people worldwide. The new company is registered in Zürich with a small headquarters there but with the top executives based in all three regional units – Vaesteraas, Baden and Mannheim.

After the merger, the Swedish Chief Executive, Mr Percy Barnevik, said that it made ABB number one in the global power business, with the only other group able to offer a complete range of power-plants being what he called 'Japan Incorporated'. (He saw Hitachi, Mitsubishi and Toshiba as a single competitor.)

In April 1988, ABB agreed to pay Westinghouse $500m to set up two joint ventures. One of these relates to the manufacture, sales and servicing of largely Brown Boveri's steam turbines and generators in North America. The other involves equipment for transmission and distribution (transformers, switchgear and control products) in the same territory. However, in February 1989, the US Justice Department objected to the steam turbine venture on the grounds that it would leave General Electric as the only other supplier in the US and required ABB to divest itself of its Wisconsin transformer business before completing the second transmission and distribution venture. As a result a revised single venture has been constructed with ABB having 45 per cent, but with a 'put and call' option on the table for ABB to take complete control if this were permitted later.

In November 1989, ABB announced it had made an agreed bid for the US company Combustion Engineering, valuing the company at $1.6bn. This was followed in December 1989 by the announcement of a joint venture – NEI ABB Gas Turbines – with NEI, a subsidiary of Rolls-Royce. The declared objective was the UK market for advanced combine cycle and cogeneration power stations expected to develop in the 1990s in the UK. (ABB has built some sixty so far of this type of power station around the world.)

ABB has also bought the steam turbine activities of the AEG, has a 40 per cent holding in Brel (British Rail Engineering Ltd) and has an agreed joint venture with Finmeccanica's Ansaldo company to restructure the Italian power generating industry. Finally, ABB now wholly owns ABB Kent (Holdings) and claims to be the second largest manufacturer of water meters in the world. It is said to hope that the UK water authorities will eventually switch to meters following the privatization of water supply in 1989.

While this dramatic pursuit of critical mass was going on, the French state-controlled Compagnie Générale d'Electricité with a controlling interest in Alsthom was discussing a merger in this sector with GEC. Alsthom is a major supplier of power generation equipment, the manufacturer of the world's most advanced railway transport (the TGV – train à grande vitesse), has a majority stake in ACEC, the Belgian energy and railway equipment group and a 45 per cent stake in the energy business of MAN in Germany.

In December 1988, the formation of GEC–Alsthom was announced, a company with over £4bn sales and some 80,000 employees. In May 1989, GEC and CGE moved closer with an allied joint venture embracing GEC's industrial controls and measurement businesses outside GEC–Alsthom but with a new subsidiary of CGE now christened CEGELEC.

GEC–Alsthom has now formed a subsidiary, named European Gas Turbine Co., in which GE, the world leader in industrial gas turbines, has a 10 per cent share. It combines GEC's former activities in small gas turbines (Ruston) with those of Alsthom's in large machines up to 212 MW and AEG's gas turbine business in Germany which GEC–Alsthom recently purchased.

It is an interesting reflection of the new competitiveness of the international market that while GEC–Alsthom have subsequently secured orders for the trains to be built for the Channel Tunnel operation, Mr Jean-Pierre Desgorges, GEC–Alsthom's Chairman and Chief Executive, has complained bitterly about the loss of anticipated UK power station orders to Siemens and ABB. So much for the old order of work being found for national industries!

The pharmaceutical industry

Without innovation the pharmaceutical industry is nowhere. There is, of course, the commodity market for 'over-the-counter' medication, but this is peripheral to the main and rapidly growing business in innovative drugs.

In pleading for rewarding prices for their successful innovations, the pharmaceutical industry has had to open up this activity to public scrutiny. Sandoz, in an example submitted to the National Economic Development Office, stated that over a period of two or three years it synthesized 10,000 compounds, only seven of which reached the stage of being evaluated on human beings and, of these, only one after twelve years of work was marketed – all at a total cost of SF 90m.

This is a far cry from the apocryphal story of Fleming's discovery of penicillin and, as a result, the world's drug business is developing its own macroindustrial structure. These are early days, however, where a new discovery can move a company rapidly up the ranking – Glaxo moved in the 1980s from a lowly twenty-fifth or so up to a position very close to number 1, powered by its drug Zantac as the world's best-selling prescription drug.

In 1989, the ranking by corporate sales in 1988 of the world's biggest prescription drug companies was as shown in Table 3.3.

Table 3.3 *The world's biggest prescription drug companies*

Company	Sales ($bn)
Merck (US)	4.9
Glaxo (UK)	4.0
Bristol Myers/Squibb (US)	3.9
Hoechst (Germany)	3.8
Bayer (Germany)	3.3
Ciba-Geigy (Switzerland)	3.1
Smith Kline/Beecham (UK/US)	3.1
Sandoz (Switzerland)	2.8
American Home Products (US)	2.7
Eli Lilley (US)	2.7
Tekeda (Japan)	2.5
Pfizer (US)	2.5

The recent Smith Kline–Beecham merger itself could be an indication of the start of a process in this sector which can be expected to follow the pattern of other innovation-based industrial sectors into the next century.

Joint ventures seem also to be developing in this sector. Merck and DuPont, the world's largest pharmaceuticals company and one of the biggest chemical companies, announced in September 1989 that they had agreed to work together on a new class of drugs to treat high blood pressure and heart trouble. In exchange, because the pioneering work was DuPont's, Merck gave DuPont exclusive rights in North America to their Parkinson's Disease drug, Sinemek, which had sales of over $100m, and marketing rights to Merck's hypertension drug in Europe – a big deal with important implications for the future of both companies.

The industry presents an extreme example of the problem of recovering the investment in an innovative product. Once launched and available for imitators to copy, the large element in the selling price required to recover the investment can only be secured in the monopoly afforded by patent protection. As the time from patenting to securing approval to market the product may be ten years or more, the time remaining before the patent expires – and the so-called generic copies come on the market – can be quite short. Intensive marketing worldwide to the medical profession and health care organizations is therefore essential – an argument for pursuing critical mass in this facet of the innovative process. Even then, with the volume maximized, the negotiations on price with the big customers – often state health services – become of paramount importance.

Telecommunications

The heretofore relatively peaceful area of telephones and electromechanical telephone exchanges has been disrupted by the impact of digital information technology. The headlines were first made by the UK GEC's hostile takeover bid for Plessey in which the subcritical mass of the UK telecommunication equipment manufacturing companies was quoted in support by GEC.

The proposed takeover was referred to the UK Monopolies and Mergers Commission, which recommended the Secretary of State to reject it (MMC, 1986). The main objection in a majority report was the impact on competition within the UK, particularly in the market for defence electronic equipment. It rejected the arguments of GEC that a larger company would be better able to compete in world markets. A dissenting view was expressed by one of the Commission's six members, Mr Baillieu.

Mr Baillieu's dissent argued that three important considerations had not been given sufficient weight in the majority conclusions:

1 The size and sophistication and essentially monopolistic powers of the two major UK customers, British Telecom and the Ministry of Defence.
2 The international dimension of the markets for both telecommunications and defence electronics where overseas competition in price and quality can no longer be ignored.
3 The size of the development effort needed for the next generation of equipment which will be beyond the capability of any one national grouping, let alone one company.

He concluded that 'to take an excessively purist line about a small loss of domestic competition is to perpetuate the balkanization of an important sector of British industry.'

However, the report did find that there would be advantages if the two companies rationalized their overlapping interests in the telecommunications market. (British Telecom had already decided to put out to tender further orders for its public digital switching requirements and had chosen the Swedish company Ericsson's System Y in preference to continuing with System X made by both GEC and Plessey.)

The report had just about been accepted by the UK Government when Groupe CGE of France and the US company ITT astounded the telecommunications world with the announcement of a deal which would create a major worldwide company by putting together their activities in this sector. The joint venture, which would put all ITT's worldwide interests in telecommunications outside the USA alongside CGE's Alcatel subsidiary, created a company with annual sales of some $10bn, operating in seventy-five countries and employing 150,000 people. As if this were not enough turmoil, another French company, CGCT (Compagnie Générale de Constructions Téléphoniques) announced it was encouraging Siemens to consider forming an association with it and other French companies in the telecommunications sector. After much jockeying for position, in which the US world market leader AT & T competed with Siemens, the Swedish company Ericsson emerged in April 1987 as the French Government's choice to take over the state-owned group in conjunction with the French company Matra.

In October 1987 GEC and Plessey announced the intention of merging their telecommunications interests to form GPT. This was completed in 1988, creating a 50:50 company with some £1.2bn sales. It was with all this – and more – in mind that Mr Karlheinz Kaske, Chief Executive of Siemens, and Lord Weinstock, Managing Director of GEC, came together and launched their combined bid for Plessey in November 1988.

In April 1989, the UK Monopolies and Mergers Commission unanimously reported that there were no reasons on the grounds of competition for blocking the bid except in defence and traffic control systems, where it recommended the UK Government to negotiate undertakings to ensure that competition was not diminished. These negotiations were completed in three months, with many of the conditions imposed in the name of UK defence interests remaining secret. However, it was announced that all the directors and top executives of Plessey defence companies acquired by Siemens or jointly by Siemens and GEC should be UK citizens. In September 1989, Plessey conceded defeat and the implementation of the merger and reconstruction began.

As if this were not enough, the Commission of the European Communities, in turning its attention to telecommunications, in a document entitled *Telecommunications: the new highways for the single European Market* (Commission of the European Communities, 1988) remarks on the fast-growing world market for telecommunications services, estimated to be worth well in excess of 500 billion ECU (1 ECU equals about 1.1 US $) and suggests that 'Europe has two options: either to participate on equal terms in this dynamic transformation of the world economy, or to become a second-rank partner and grow steadily poorer'. The document further states:

> Today national markets in the Community are no longer big enough to cover the ever-increasing costs of research and development. The growth of those costs is such that software has become much more expensive than hardware. It is estimated that at the beginning of the 1990s the software for a public switching system, which will incorporate about three million programmed instructions (compared with one million at present), will represent 80 per cent of the total cost of the system; in 1970 the proportion was 20 per cent. The research and development cost of such a system is now 1,000 million ECU, where the electromechanical systems of 1970, which had a useful life two or

three times longer, cost only 15 to 20 million ECU. Investment on this scale cannot be undertaken unless one can count on obtaining 8 per cent of the world market, but none of the national markets in the Community amounts to more than 6 per cent (while Japan has 11 per cent and the United States 35 per cent).

It is, therefore, no longer possible to develop European telecommunications according to the traditional model of 'national champions' which has led to the development of eight different types of digital switching systems in Europe, compared with two in Japan and three in the United States. Europe can no longer afford the luxury of costly duplication of research and development within national markets that are partitioned from each other. It cannot allow continuation of the equally serious delays and shortcomings in the development of new high value-added services, due to divergences between the timetables and standards of the various Member States.

It goes on to describe the challenge the Community faces in this respect, all part of a wider issue that we return to in Chapter 7 on the European Community policy for innovation. Formation of alliances in the communication industry (some illustrated in Figure 3.4) has followed a

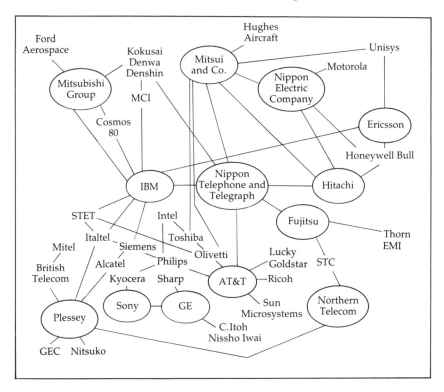

Figure 3.4 *Alliances in the telecommunications industry*
Source: Devlin and Bleakley (1988)

similar pattern to that experienced in the automible industry (Figure 3.3), involving cross-national and intercontinental grouping (Devlin and Bleakley, 1988).

Final remarks

No more authoritative statements on the subject of critical mass can be found than those contained in signed articles on this subject by Lord Weinstock of GEC and Mr Karlheinz Kaske of Siemens. Provoked no doubt by the public debate on their joint assault on what, by UK standards, was a big and established member of the upper echelons of British industry, they unequivocally declared their concern about the adequacy of the scale of GEC (£6bn sales) and Siemens (£19bn sales) to compete successfully in the global market for certain innovative products. Under a general caption 'The Importance of Critical Mass' these extracts appeared in the *Financial Times* in April 1989:

> *Lord Weinstock* – The capacity to spend on R&D in this game is largely determined by turnover, and turnover is determined by markets. GPT alone does not now have the markets or turnover to justify the spend necessary to develop the successor to System X. GPT and Siemens together do, although even GPT and Siemens operating a joint systems development programme will be smaller than the larger Japanese and American telecommunications businesses.... The British view has been that GEC is too big. The world perception is that GEC is not big enough ... the minimum scale for effective survival is always rising. A niche can easily become a tomb (Weinstock, 1989).
>
> *Mr Karlheinz Kaske* – At the root of these changes, driven by the accelerating pace of technological advances, is the enormous increase in research and development expenditure which is required to develop a new generation of electronic systems, together with the growing need for substantial investment in new manufacturing technologies and equipment. To amortize these costs, increasingly large production and sales volumes are required.... Whereas the development of a traditional electromechanical public switch (telephone exchange) cost about $100m, the development cost of one of the digital switches in current use is about $1bn. Future systems are expected to cost up to $2bn.... For the next generation, probably only the top half dozen of companies in the world will have sufficient volume and access to world markets (Kaske, 1989).

This is not the pursuit of size for size's sake – not 'folie de grandeur' – but a declaration, driven by fear, that without critical mass the survival of the enterprise would be in jeopardy.

Size, as many writers on management have pointed out, has its own problems and disadvantages. In Chapter 4 we turn to examine 'bigness' in its broader aspects.

References

Commission of the European Communities (1988), *Telecommunication: the new highways for the single European market*.

Devlin G. and Bleakley M. (1988), 'Strategic Alliances – Guidelines for Success', *Long Range Planning*, vol. 21, no. 5, pp. 18–23.

Kaske, K. (1989), 'The Importance of Critical Mass', *Financial Times*, 7 April.

MMC (1986), *General Electric Company plc and the Plessey Company plc: a report on the proposed merger*. HMSO, London.

Times (1989–90), *Times 1000, 1989–1990 edition*, Times Books.

Weinstock, A. (1989), 'The Importance of Critical Mass', *Financial Times*, 7 April.

4 Big is necessary

- **Schumacher's thesis**
- **Failure of large systems**
- **Effect on the environment**
- **Distribution of company size**
- **Advantages of scale and size**
- **Theoretical considerations**
- **The unit cost model**
- **A linear cost function**
- **Empirical evidence**
- **Economies of scale**
- **Dean's empirical studies**
- **Effect of output on unit cost**
- **Conclusion**

As business becomes increasingly international and as innovation policy becomes more important to the long-term strategy of an enterprise, the question of size needs to be addressed: How big should an enterprise be in order to have the necessary resources to withstand global competition? As argued in the previous chapters, critical mass is a central strategic issue for technologically-based industries and many commentators have, consequently, strongly advocated that bigness is not only necessary but inevitable; others, however, have been staunch supporters of the small enterprise as the model of vitality and efficiency. This chapter continues to consider the economic aspect of size and its effect on unit cost. But first let us examine in some detail the arguments for keeping the enterprise small.

Schumacher's thesis

In the early 1970s Schumacher published his seminal book *Small is Beautiful*, in which he set out to challenge the relentless march of large organizations 'towards vastness' and to highlight the consequences of their increasing dominance over our economic system (Schumacher, 1973). He was particularly alarmed about the detrimental effect of corporate size on individuals within these organizations: 'Most of the sociologists and psychologists insistently warn us of its inherent dangers – dangers to the integrity of the individual where he feels as nothing more than a small cog in a vast machine and where the human relationships of his daily working life become increasingly dehumanized.'

This theme has since been taken up by many writers – sociologists, economists and politicians – who have become very concerned about the dominating power that large organizations can exercise and the degree to which they can become immune to external influences, while pursuing strategies aimed at advancing their own corporate needs. The fears of the rising power of large corporations can be summed up under several headings.

First, there is the fear that corporate objectives are primarily set with the view of improving the financial performance of the corporation, of furthering the status and comfort of key executives and of generating good dividends and future prospects to shareholders, while ignoring specific needs of most individuals within the organization or those of the community. There is an increasing feeling of resentment that those in charge of large corporations have an inordinate influence on the lives of ordinary citizens, and that they do so without being elected and without being openly accountable to the public for their actions.

Admittedly, small companies are not directly accountable either, but their impact on the local community or on the national economy is minuscule by comparison. In contrast, if a large company decides to close a plant employing several thousand people, or to move it elsewhere, the consequences for the local community can be catastrophic, not just for those losing their jobs in the plant, but for suppliers, for all businesses in the locality, for shops in the high street, for professional practices, for prices of real estate, for future development prospects, and even for the maintenance of many social amenities. One decision by a major employer can convert prosperity into blight, and there is little that the community, or those affected, can do about it.

Second, large organizations tend to become bureaucratic, cumbersome to deal with and difficult to manage. As a result, personal relations are destroyed, or become, as Schumacher put it, 'dehumanized'. The individual employee feels lost in a vast organization, his needs and aspirations seem to be ignored, his contribution to the wellbeing of the organization is seen, at best, as marginal; his efforts are probably recognized only by the few people around him, and he is likely to be quite unknown to the senior executives who wield the real power. It often makes little difference whether he works hard or whether he simply passes the time. A risk-averse mentality becomes rife among cohorts of organization-men, who need to conform without ever being seen to have made mistakes. The large corporation can become a congenial habitat for faceless bureaucrats, an elaborate edifice of rules and regulations, a massive freighter with enormous inertia that impedes any attempt to change course. This vision of the organization is shared by individuals outside who have dealings with it as customers, as suppliers, or simply as members of the general public.

Third, there is a fear of the oligopolistic, and in some cases even monopolistic, power of the large corporation. It can drive small competitors out of business by indulging in price wars, which small businesses can ill-afford, or simply by taking them over. It can dictate the price in the market by controlling the level of supply; it can decide on the range of products that may be made available in the market, and determine their specifications, the rate at which new models are issued, the distribution network, the arrangements for after-sale service and the conditions under which it is provided. It can dictate terms to suppliers, who may become so dependent on their major or only customer, that their profitability, or even their very survival, may be continuously in the balance.

The fear of oligopolistic or monopolistic power stems from the realization that it can stifle competition and reduce customer choice, while at the same time be able to erect what seem insurmountable entry barriers to new entrepreneurs anxious to challenge the status quo. It has often been asserted that in a free economy large corporations would have tended to abuse their power in this way, had it not been for the restraining influences of anti-trust legislation and the activities of various agencies set up to protect the consumer and 'the public interest'. This is why even governments totally dedicated to free enterprise and the principles of a market economy, take seriously the possible dangers lurking behind monopolies and try to ensure that customers do have a choice.

Large corporations do not necessarily confine their activities to a single industry, and some have developed into conglomerates with many interests, but concern over their power remains. As Anthony Sampson states in his book on ITT (Sampson, 1973):

> The case against conglomerates was not indeed an easy one to make, for by their diversity they cut across the old definitions of monopoly and restriction of trade; and it was partly to avoid anti-trust action that they had become diversified. Few of them dominated a single industry, and theoretically world industry might be run by a handful of vast conglomerates, each competing with scores of industries, without cutting across anti-trust laws. The fundamental argument against them was not so much economic, as political and social: that they restricted individuality and freedom of choice, that they centralized and concentrated activities which could survive separately, and that they were simply, in a word, too big, too ubiquitous and too powerful.

Failure of large systems

Apart from questions relating to market dominance and abuse of power, the issue of size per se has a further serious implication. As stated elsewhere (Eilon, 1989a):

> There is another problem that has perhaps not been sufficiently ventilated, and that concerns the possible dangers to society when a large and complex system, like one of the large corporations, simply ceases to function, not necessarily because of evil or sinister intentions of their leaders, not because their goals could be in conflict with national interests (although such circumstances do often arise), but because of technical malfunctioning of the system. Now, it may be argued that such a catastrophe is just not likely to happen, that a large corporation such as IBM or Unilever or ITT is too robust to be vulnerable either to outside disturbances or to internal failures of their control systems. Yet, large-scale and complex systems do fail, and they do go astray. The failure of the electric supply system on the eastern seaboard in the United States in 1967 was totally unexpected, so was the sudden collapse of Penn Central in 1970.

There can inevitably be far-reaching economic consequences to a locality, or even to a whole country, when a large corporation experiences financial difficulties or an organizational collapse (the discussion on 'betting the company' in Chapters 1 and 2 is also relevant in this context) and when difficult decisions are made at a remote head office to the distress of local government trying desperately to cope with the consequences (Eilon, 1979b).

The development of ever larger computer systems in recent years, with mind-boggling complexity, designed to control the activities of large corporations, public utilities, and government administration, let alone defence establishments, makes a breakdown quite plausible, irrespective of the increasing sophistication of safety precautions built into these systems. One example of this was the big crash of the stock market in October 1987, which was exacerbated by programmed trading, namely by a wave of automatic selling triggered off by computer programs when prices dropped below pre-specified thresholds, and curiously enough the stock market witnessed another mini-crash almost exactly two years later, when failure to proceed with a financing deal for the management buyout of United Airlines suddenly triggered off an uncontrolled wave of selling, which fortunately fell short of the 1987 collapse.

Even if steps are taken to ensure that the specific cause of a malfunction can be identified and dealt with in time, to ensure that such a calamity does not happen again, it is important to appreciate that many preventive measures can only be taken after disasters have already occurred and after their causes have become fully understood. But all the possible reasons for system breakdowns cannot be fully anticipated and new circumstances may arise, which may generate new causes for malfunction and disaster. In addition, large systems become vulnerable to sabotage, spurred on by political motives, or even perpetrated by wilful competitors. As a system increases in size and complexity, so may the dire consequences of possible failure, innocent or malevolent in origin, and the ripple effect on the national and even the international economy can be far reaching.

Effect on the environment

And there is, of course, the effect on the environment. It is now fashionable to be 'green' and to pontificate on the debilitating effect of industry on the future of the globe. This is, in fact, not a new issue. Schumacher mentioned the effect on the environment as one of his major concerns about big companies (Schumacher, 1973). If a small enterprise misbehaves, for example if it pollutes the air or the local river, its activities are more amenable to inspection and control, whereas a big corporation often has sufficient clout to resist what it may regard as interference in its affairs, or it may resort to elaborate delaying tactics in an effort to save money, while pollution continues unabated. But apart from the effect on the environment from daily operational activities, there is the possible risk of accidents, which can be devastating when they happen on a large scale. There have been, alas, many examples of such accidents in recent years: large tankers breaking up and causing pollution to beaches and destroying wildlife, crashes of jumbo jets with many lives lost, explosions of chemical plants with horrendous loss of life and devastation to local communities (such as the plant at Bhopal in India), and so on. The environmental risks, therefore, from large systems, which are usually controlled by big corporations, are much greater than when small firms are involved, both from daily operations and from unexpected accidents.

These issues are discussed by Dathe (1982):

> Today, problems of scale are an important issue for people who study the economic and societal causes of our industrial development. A supertanker that is sized to fulfil the requirements of economic transportation may become a 'super offence' to coast regions in case of an accident. The centralization of the electric power supply, which has definite advantages in balancing partial malfunctions in routine service, may become the cause of a major breakdown of large network areas in the case of a sudden cumulative overload. [Consequently,] nobody should be surprised that – in spite of pronounced gains in efficiency that are achievable only by large technical units – many people are reluctant to believe in the advantages offered by sheer technical magnitude.

In addition, high levels of unemployment in Europe and the US during the 1970s and 1980s became the most pressing problem on the political agenda. Against the background of industrial rationalization, which saw many big companies shedding labour in vast numbers, it became clear that big companies were not likely to solve the unemployment problem and that, instead, salvation lay with the very large number of small companies, which form the backbone of the industrial economy of any nation. It was argued by many politicians and economic commentators in Britain that if every small company (employing, say, less than twenty people) were to take on one or two additional employees, the scourge of unemployment could almost be eradicated. Encouragement for small businesses to expand, and for new businesses to be set up, has become a political priority in many countries.

For all these reasons, 'small is beautiful' is a slogan that has won wide appeal. But is it? If small is beautiful, why is it that big companies continue to dominate the industrial scene and that over the years (leaving aside blips of economic upheaval) they tend to become bigger and bigger? We need to review some of the theoretical arguments for and against economies of scale and examine some of the empirical evidence in order to throw additional light on this issue.

Distribution of company size

A Pareto curve constructed for any economy reveals that industry consists of a comparatively large number of small firms. In many countries more than 80 per cent of all the firms have less than 100 employees each. But such a statistic conceals the power and influence that large corporations have on the national economy, even when they are relatively few in number. Large corporations, like General Motors, IBM, BP, Shell, Unilever and many others, already have annual incomes that exceed the national budgets of small countries, and some continue to expand through organic growth and acquisitions. Their vast assets and sheer momentum give them a protective layer, immunizing them from external shocks and ensuring (barring exceptional political upheavals) their continued survival and prosperity, so that most of the large corporations are associated with a sense of robustness and dependability. All these arguments suggest that, in spite of what Schumacher and his followers preach, there are good reasons for many firms to conclude that in reality big is necessary, if not beautiful, and that this is why big corporations strive to become even bigger. We shall now further examine some of the reasons for this trend.

Advantages of scale and size

Gold has conducted extensive research in many industries on the question of scale. He says (Gold, 1982):

> Increasing competitive pressures in domestic and international markets have stimulated efforts in many industries to gain what are widely believed to be advantages of 'scale economies' through the building of progressively larger operating units. Such tendencies are apparent in broad sectors of manufacturing – including chemicals, steel, pulp and paper and cement – as well as power generation, mining, shipping and agriculture. This reflects the spread of faith in the benefits of scale increases beyond engineers and industrial managements to governments, which then foster larger operations in the hope of strengthening the competitive position of their industries.

The widely-used term 'economies of scale' has come to signify the advantage of spreading fixed and indirect costs over a larger volume of output and thereby achieving lower unit costs. But the term is somewhat imprecise. It does not indicate whether the advantage is derived from generating a bigger output on a given plant, or from the use of a bigger plant, or from the efficiency that technology can offer the bigger plant. In this context, Gold makes an important distinction between scale and size. Scale refers to the capacity of the operating unit, whereas size reflects the total output. Thus, the output of a given plant may be doubled, either by erecting another production unit, so that two identical units would be operating side by side, or by replacing an existing unit with one having twice its original capacity. The former alternative results in doubling the size but not the scale of the plant, whereas the latter alternative results in both. Usually, an increase in scale automatically means an increase in size, though in some cases an increase in scale only provides the scope for increasing output through increase in *available* capacity, but if this capacity is not fully utilized, then output (and hence size) may remain unchanged.

A decision to increase scale may be motivated by one or a combination of the following reasons:

1 To meet an increase in demand, either in response to an expanding market, or as part of a strategy to increase market share

When an existing plant is used to its maximum or near-maximum capacity, pressure usually mounts to increase capacity by adding a further production unit, or by replacing an existing unit with one of larger capacity.

2 To be able to meet sporadic or seasonal demand at short notice

A plant working continuously at full capacity would not be able to meet sudden upsurges in demand. A familiar example is that of beer and soft drinks, where fluctuations in demand are affected by the weather and are superimposed on a seasonal pattern. Having excess capacity, over and above the regular constant demand level, means that a risk is incurred of having to contend with underutilized capacity, sometimes for extended periods. However, when an upsurge in demand does occur, the excess capacity is immediately put to good use and can be very profitable.

3 To exploit up-to-date technology

In some industries rapid progress in technology accelerates the rate of obsolescence of existing plant and machinery, and pressure to modernize becomes inevitable, particularly when competitors invest in new technology. Proposals for the replacement of existing plant are automatically accompanied by an assessment of market prospects, and if they justify an investment in a new plant, the temptation to opt for a larger scale is almost irresistible. The promise of new technology is that it would yield:

- Better quality in manufacture, reducing defects and reprocessing.
- Improved facility to introduce new designs.
- More flexibility in manufacture, particularly in reducing set-up times and costs when switching from one product to another.
- Greater flexibility in the introduction of new designs.
- Reduced need for plant maintenance, as well as a lower incidence and severity of breakdowns.
- Better control of production and better integration of individual operations.
- Higher labour productivity (i.e. increase in output per man).
- Opportunity to become less dependent on scarce inputs, such as highly-skilled labour (e.g. through automation), or rare materials.
- Higher materials productivity (by reducing the amount of scrap).
- Savings in the use of other expensive inputs, such as energy.
- Improved safety and environmental standards.
- Shorter production operations and overall production times.
- Better utilization of space and improved materials handling.
- Lower inventories of incoming materials and of work in progress.

The motives for introducing new technology are, therefore, wide ranging. Each of the objectives listed above would be sufficiently compelling, given the right circumstances, to justify investment in a new plant. Some of the objectives involve benefits of a qualitative nature (such as improvement in control or the working ambience), others may relate to new legislation (for example, on air pollution standards). In the main, though, the ultimate aim of most of the above objectives is a reduction in unit cost, which is one of the central ingredients of a corporate strategy designed to achieve a competitive advantage in the marketplace and which, in turn, helps to increase profit and cover the launch cost of a new product (as discussed in Chapter 1).

Theoretical considerations

One of the main issues considered in evaluating the competitiveness of a product is that of unit cost. Texts on managerial economics (see, for example, Pappas and Brigham, 1979) make a distinction between short-run and long-run cost functions which describe the relationship between cost and volume (or rate of output). The short term is defined as a period in which all the operating conditions and all the inputs (except those that directly depend on the output, such as materials and direct labour) are fixed. As Dean puts it (1976, p. 4):

> [The] theoretical production function is a rigorously specified relationship between the factory's rate of output and the rates of input of productive services (machine hours, materials, labor hours, etc.), all measured in physical terms. It assumes that management is perfect, i.e. that executives will always achieve the maximum rate of output that is technologically possible with the specified rates of use of the productive services. It also assumes that those inputs which are not fixed in total can be varied continuously (i.e. by tiny increments) either individually or together, and that the resulting output will also increase smoothly by tiny increments.

In discussing cost functions it is important to emphasize that we are not merely concerned with production costs, but with total costs, including the cost of innovation and launch, as implied in the discussion in Chapter 3. Second, we need to make a distinction between fixed costs and variable costs. Theoretically, fixed costs are those that do not change with the level of output, namely they remain unchanged when the volume increases, and similarly they cannot be reduced when the plant faces a fall in market demand. Such definitions are said to apply for short-term cost functions, since in the short term management is constrained in what can be done to change the fixed cost element of the total cost. In the long run, however, management can alter all the factors that affect production, including all the fixed elements, the machinery and plant, the design and mix of products, the production methods and the operating conditions. In short, it is generally assumed that management is able to adjust all its cost determinants in the long term to suit the level of activity expected of the plant, including the use of specialized facilities, and thereby optimize operations to achieve the best possible results. The relationship between short-run and long-run cost curves is discussed further below.

The unit cost model

A typical short-term cost curve is shown in Figure 4.1, where:

F = the fixed costs, assumed constant throughout the volume range; for the purpose of this discussion it is important to appreciate that fixed costs are not confined to operating costs, but include the important elements of innovation, product launch and investment in fixed facilities (plant and machinery), depreciated over the life of the product.

S = the variable costs, which start at 0 (no variable costs for no output)

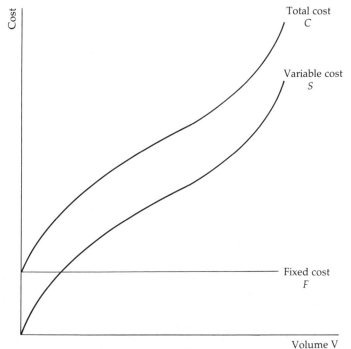

Figure 4.1 *Short-term variable cost and total cost curves*

and increase with volume; typically, it is assumed that the variable cost curve, based on constant input factor prices, increases first at a decreasing rate up to a given output level and then at an increasing rate, thereby producing an inverted S-shaped curve for the variable costs which suggests that the marginal productivity of the variable production inputs initially increases and then declines. According to Pappas and Brigham (1979). 'This relationship is not unexpected. A firm's fixed factors, its plant and equipment, are designed to operate at some specific production level. Operating below that output level requires input combinations in which the fixed factors are underutilized, so output can be increased more than proportionately to increases in variable inputs. At higher than planned output levels, however, the fixed factors are being overutilized, the law of diminishing returns takes over, and a given percentage increase in the variable inputs will result in a smaller relative increase in output.'

C = the total cost, obtained as the sum of the other two cost elements, namely:

$$C = F + S$$

If the total cost C is divided by the level of output volume V we derive the unit cost c, i.e.

$$c = C/V$$

The unit fixed cost (i.e. fixed cost per unit) is F/V and the average variable cost per unit is S/V, the total average unit cost being the sum of the two, namely:

$$c = F/V + S/V$$

and shown in Figure 4.2 as a U-shaped function, which has a point of minimum unit costs, corresponding to the 'optimal' volume for this plant. Note that the U-shaped curve for the total unit cost in Figure 4.2 is the direct outcome of the assumption of the inverted S-shaped curve in Figure 4.1. If the variable (or marginal) cost per unit were assumed to remain constant, irrespective of the output level, then the total unit cost would be a continuously declining function as the volume increases. We shall return to discuss some of the implications of such an assumption later.

Figure 4.2 shows the unit cost function for an existing plant and for existing technology. A series of different plants is shown in Figure 4.3 each being U-shaped with its own point of minimum unit cost. Management can then decide which plant to choose for any level of output envisaged for future operations. Although the plants shown in Figure 4.3 constitute a discrete set, it is postulated that a long-run cost curve can be

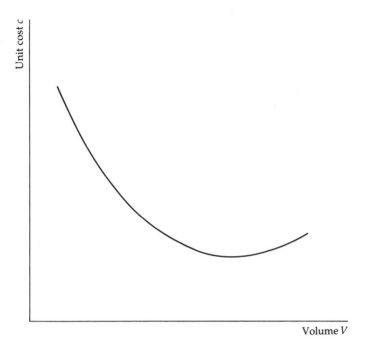

Figure 4.2 *A 'typical' short-term U-shaped curve for unit cost*

constructed as a continuous envelope of the short-term curves, and this is shown in Figure 4.4. All these cost functions indicate an initial decrease in unit cost as volume rises, until a minimum point is reached, beyond which unit cost increases. In other words, all these cost functions, both

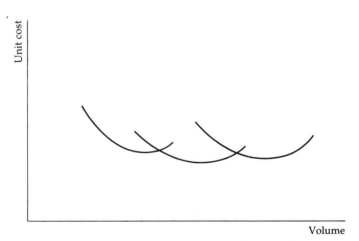

Figure 4.3 *'Typical' short-term unit cost curves for a series of plants*

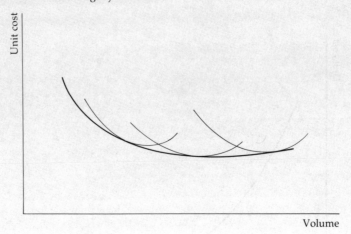

Figure 4.4 *A long-term unit cost curve constructed as an envelope of a series of short-term cost curves*

short term and long term, are assumed to be U-shaped, and the optimum scale of plant is then determined from the long-term curve, such as the one in Figure 4.4. Up to the optimal point the long-run average unit cost declines and production is said to display economies of scale, while beyond that point diseconomies of scale are manifest.

Economies of scale result from a combination of many factors, chief amongst them being: spread of fixed costs over a larger volume output, greater labour productivity achieved in a larger plant through a greater degree of specialization, use of more special-purpose equipment coupled with more efficient and more productive use of the equipment, better organization and control of production and inventories, use of quantity discounts in bulk purchasing of materials and services, and better utilization of space and transport facilities.

As for the phenomenon of diseconomies of scale, the usual argument is (Pappas and Brigham, 1979):

> At some output level economies of scale typically no longer hold, and average costs begin to rise. Increasing average costs at high output levels are often attributed to limitation in the ability of management to coordinate an organization after it reaches a very large size. This means both that staffs tend to grow more than proportionately with output, causing unit costs to rise, and that managements become less efficient as size increases, again raising the cost of producing a product ... While the existence of such diseconomies of scale is disputed by some researchers, the evidence indicates that diseconomies may be significant in certain industries.

For a multiplant firm, where the total cost is the sum of the costs of the individual plants, it is then suggested (Pappas and Brigham, 1979) that the

Big is necessary   75

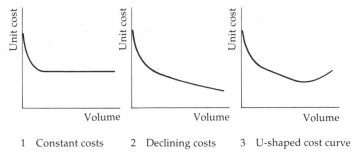

Figure 4.5 *Three possible long-term unit cost curves*
Source: Pappas and Brigham, 1979

resultant long-run unit cost may take one of the forms shown in Figure 4.5:

1 The costs decline and then remain constant and unaffected by changes in volume.
2 Economies of scale continue to operate and there is no optimal level of output.
3 The U-shaped curve is manifest again, as in the case of long-run costs in a single plant.

As to whether management would be wise to select the lowest unit cost solution on the long-term unit cost function, Pappas and Brigham add a caveat that the optimal point is valid for the case of a constant level of output, otherwise the decision would have to take account of the variability in demand. The example given is shown in Figure 4.6, where two

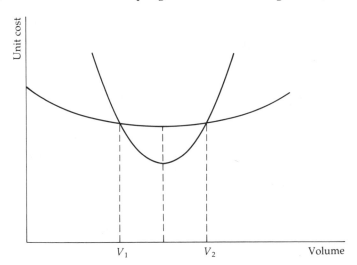

Figure 4.6 *An example of two alternative plants having the same optimal point*
Source: Pappas and Brigham, 1979

plants A and B have their minimum unit cost at the same output level, but because plant A is more specialized, its cost function is much steeper on either side of that point compared with plant B. If the level of demand is certain to fall between V_1 and V_2, then plant A would obviously achieve lower unit costs, but plant B is superior for demand that falls outside this range. The analysis must, therefore, be based on the expected unit cost, i.e. when the probability distribution of the demand (or output level) is taken into account. This problem is further exacerbated when demand is highly volatile and when the alternative plants do not have the same points of minima with respect to volume, and this is shown in Figure 4.7.

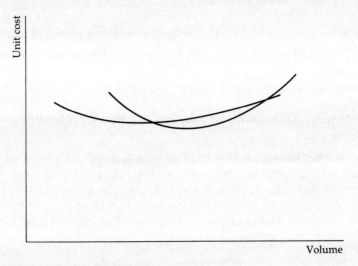

Figure 4.7 *Two plants with different optimal points*

A linear cost function

A special case of the total cost curve is a linear function, such as the one shown in Figure 4.8, where the fixed cost F is assumed constant and the variable cost S rises linearly with volume V, so that $S = sV$, with the marginal cost s remaining constant throughout the range. The total cost then becomes:

$C = sV + F$

and the unit cost:

$c = s + F/V$

then declines continuously as the volume V increases.

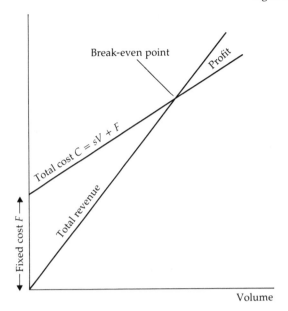

Figure 4.8 *The break-even point*

This cost function is of particular interest for several reasons:

1 It is very convenient for analytical purposes, such as break-even analysis (Pappas and Brigham, 1979), typically depicted in Figure 4.8, so that the total cost, the unit cost and the expected profit can be easily computed for any level of volume.
2 Empirical data, on which cost functions have to be postulated, almost invariably involve a scatter of points derived from historical data, which involve different levels of activity at different times, without the assurance that all the operating conditions (other than volume) were absolutely identical. Fitting a curve through a given set of points may take several alternative forms, depending on the amount of data available and on the degree of scatter. Linear functions are often as convincing a fit as any elaborate multiterm power functions. For example, the empirical results shown in Figure 4.9 may suggest a flat inverted S function, but many will regard a linear regression as a sensible alternative. Unless there are good reasons to insist on a more complex functional form, a linear fit is generally thought to be appropriate for most practical purposes.

Empirical evidence

It is clear, therefore, that if the unit cost function takes the U-shaped form shown in Figure 4.2, then there are no economies of scale beyond a certain

78 The Global Challenge of Innovation

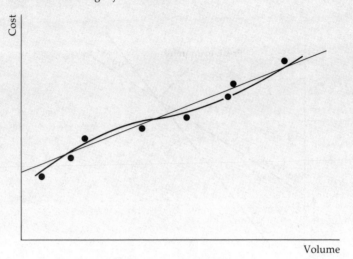

Figure 4.9 *Fitting a cost curve to available data*

level of output, whereas the prospect of a continuously declining unit cost (such as the one obtained for the linear cost function) is an obvious incentive for management to increase volume. This issue is of particular importance in view of the need for critical mass, triggered off by the high level of expenditure on new product launch, discussed in Chapter 3. A U-shaped cost curve would run counter to the pursuit of increased volume.

But what happens in reality? Are U-shaped cost curves a universal presentation of how the unit cost function behaves in practice? As Gold (1975) puts it: 'Continuing reliance on convenient assumptions in place of exploring the realities of industrial practice rendered traditional approaches to scale economics widely inapplicable in concept and all but trivial in their posited effects.'

We need, therefore, to turn to the question of empirical evidence in order to ascertain the effect of an increase in plant size, an increase in output rate and the introduction of new technology on plant performance. One of the problems of seeking empirical evidence is that it has to be based on collecting data which correspond to identical operating conditions, except for one variable, so that the effect of that variable can then be ascertained. Thus, to discover the effect of the three determinants discussed above, it would be necessary to design the empirical studies as follows:

- *The effect of plant size* Identify several plants of different capacity producing the same output under the same operating practices and using the same inputs.

- *The effect of output rate* Identify a plant engaged in different rates of production (for example when monthly output is dictated by varying market demand) under the same operating conditions. (Alternatively, compare the performance of several identical plants producing at different rates. However, the probability of identifying a set of plants that meet these conditions is rather low.)
- *The effect of technology* Identify several plants of equal capacity and equal output rate, requiring the same inputs but employing different technologies.

These brief specifications give an indication of what is required in order to determine the effect of one determinant, 'all other things being equal', to use a popular economist's phrase. Clearly, very few empirical studies can meet these stringent conditions, as 'all other things' are never equal. Take the case of plant size: It is very rare to find a wide range of plant capacities based on exactly the same technology, since the construction of a new plant of greater capacity gives the designer an opportunity to incorporate the latest technology appropriate to the envisaged scale of production, so that plant size and technology can hardly be regarded as totally independent variables. Similarly, it is very difficult to separate plant size from output rate, since the purpose of building larger plants is to provide the capability to increase output. As for the effect of technology, there is the added difficulty that – unlike plant capacity and output – it does not have attributes that can be measured on a continuous scale. It is not surprising, therefore, that convincing empirical studies are few and far between, and even then the results would be valid only for the specific case studied, rather than being conclusive for a whole industry or across several industries.

The need for statistical studies also raises the thorny problem of sample size. Only few countries have plants with a wide spectrum of size, output and technology in the same industry. Inevitably, research workers have been attracted to the idea of widening the scope of their investigations through international comparisons, which – so it is argued – would not only provide useful insights into the relative industrial efficiency and performance in different countries, but would help to increase the database for analysing the effects of scale, size and technology.

Alas, this proposition is fraught with many difficulties. If two plants of identical size and output in two countries produce an identical product, in terms of specifications and quality, at different costs, we may conclude that one is more efficient than the other (assuming the currency exchange rates and factor input prices are not artificially set). But if the plants are not equal in size or output, how can we judge whether the difference in unit cost is due to the effect of scale (or rate of output) or simply to the difference in industrial efficiency prevailing in the two countries?

In addition, it is imperative in such studies to use the same definitions and methods for measuring the variables in all the plants in the sample. This is where many studies involving international comparisons often come to grief: they tend to be conducted by different people in different countries using different definitions and measurement criteria. It is notoriously difficult in any organized (or disorganized) activity, akin to the tower of Babel, to reduce the cacophony of sound to a coherent exchange of information, and in many international studies doubt remains about comparing like with like. This is why any research programme in this field has a better chance of producing convincing results when the vast number of variables and the great complexity of the systems under study can be reduced to the bare minimum.

Economies of scale

It follows from the above arguments that there are many difficulties in determining the effect of scale on unit cost, since production cost is a function of many variables, of which the scale of the plant, measured by its capacity, is only one. The other important variables, already alluded to, are: the output rate; the technology used; the mix of inputs and their factor prices; the product specifications; the product mix (for plants producing multiproducts); the production and managerial control methods; the levels of skills of the workforce; and the many other variables loosely grouped under the term 'operating practices'.

One way of attempting to tackle this complex problem is to choose an industry which is capital-intensive and which is concerned with a single product and a very limited set of inputs, thereby reducing the number of variables. Several plants of different capacity can then be compared in order to ascertain whether economies of scale exist. Three examples of such industries are cited below:

1 The first is an old study by Edwards and Townsend (1959) relating to the capital cost of low pressure gas storage holders, summarized in Table 4.1. The range of holders included in the study is 50:1 and the capital cost per unit of capacity is taken as 100 for the smallest holder (of 10,000 cubic feet). The capital cost required declines rapidly as the capacity increases and for the largest holder in the range the capital cost drops to a ninth of that needed at the lower end of the range.
2 The second is a study of steel plants, shown in Table 4.2, where Rosegger (1975) shows that both for the basic oxygen process and for electric furnaces the investment cost declines appreciably with scale.
3 The third is a study by Bolotryi and Itin (Buzacott et al, 1982) relating to the performance of thermal power stations and summarized in Table 4.3.

Table 4.1 *Capital cost of low pressure gas holders per unit of capacity*

Holder capacity ('000 cubic feet)	Capital cost/unit
10	100
20	64.2
40	54.4
60	45.9
100	32.8
250	21.3
500	16.4
2250	14.8
5000	11.5

10,000 capacity holder taken as 100
Adapted from Edwards and Townsend, 1959

As these results suggest, capital investment, the amount of space required for the plant and the construction cost all decline as capacity increases. The number of people needed to operate the plant falls dramatically, particularly for high capacity levels, and this is reflected in a reduction in production costs (see also Betts, 1982).

Studies of various processing industries yield similar qualitative results. Although such studies do not isolate scale as the only variable (since, as we said earlier, in most comparative studies of this kind the technology prevailing in different plants is not the same, nor is the level of plant utilization and many other variables), the general conclusion is that scale (combined with the technology that is appropriate to any given plant size) is a significant determinant of costs.

Table 4.2 *Decrease in investment cost per unit of capacity for increase in plant size*

Capacity increase (in '000 of tons p.a.)	Decrease in investment (in %)	
	Basic oxygen process	Electric furnace
From 100 to 300	17	13–18
From 300 to 500	16	15
From 500 to 1000	22–28	25–27
From 1000 to 1500	12–15	12–16
From 1500 to 2000	3–9	–

Source: Rosegger, 1975

82   The Global Challenge of Innovation

Table 4.3   *Performance of thermal power stations*

	Capacity (MW)				
	200	300	600	1200	2400
Capital investment/MW	100	86	75	66	60
Main building volume/MW	100	88	84	58	51
Construction costs/MW	100	96	90	76	68
Operating personnel/MW	100	84	60	32	24
Energy production costs	100	91	87	78	70

200 MW taken as 100
Source: Buzacott et al, 1982

Another area of interest is that of air transportation, which, as Dathe (1982) puts it:

> proves to be a field well suited to the study of technological, economic, and environmental problems of unit scale. The development until 1970 was characterized by a steady increase of aircraft size which was unquestionably caused by the fact that the economically and technologically justified maximum unit scale was not reached before. Fortunately for aviation, the Boeing 747 'Jumbo Jet', the biggest aircraft, became a success after initial difficulties. As a consequence, so-called wide-body aircraft are being introduced in growing numbers both in the long-range and in the short- to medium-range market.

The use of large aircraft means that it would be possible to double air traffic volume while increasing the number of aircraft by only 26 per cent. Studies of passenger air safety, measured in fatalities per 100m passengers, show a distinct decline over a period of twenty years from the late 1950s.

Dean's empirical studies

The problem of carrying out empirical cost studies on the effect of scale and size is widely discussed in the literature, and Joel Dean, who has conducted such studies over many years, puts it succinctly as follows (Dean, 1976):

> Cost has many determinants. The two that have been most important in economic theory are the rate of output and size of plant (or firm). Others have, of course, been recognized by economists, but they have usually been viewed as contaminants – forces to assume constant in order to examine the two critical cost determinants: output rate and plant size.

He realized that conducting empirical tests across the whole of industry would not be particularly illuminating, since each industrial sector has its own range of determinants, which may not be applicable to others. He then set about to examine empirical evidence in several industries, and two series of studies are of particular interest, namely those where he attempted to determine the relationships between:

- *Cost and output rate* (in a furniture factory, a hosiery mill, a belt shop and a department store).
- *Cost and plant size* (in a finance chain and a shoe chain).

Dean carried out other studies to establish the effect of plant location in a drug chain and the relationship between cost and real profit in an electronics factory and a machinery manufacturer, but these are not relevant to our discussion.

On the effect of the output rate, Dean summarizes the problems encountered in his studies as follows:

> Statistical cost determination can be of two kinds: (1) simultaneous observation of costs of different (but similar) plants, operating at different rates of utilization, or (2) sequential observation of costs of the same plant over a period of time when it operates at different rates of output. To find a large number of plants that are sufficiently similar in equipment, management methods, and records (so as to proxy an identical plant) but that differ over a wide range of use is not easy. Nor is it easy to find a factory whose size, technology, and management methods have remained substantially constant over a period during which output rate has fluctuated widely.

The results of Dean's empirical studies were illuminating: At each of the three factories (furniture, hosiery and belt shop) 'total cost rose in a straight line as a function of the rate of output over the range of output observed. Hence marginal production cost was constant.' This meant that in each of these cases the average unit cost continued to decline as production rate increased. At the department store, Dean examined three departments – shoes, hosiery and coats – and for the first two he found that the total cost function was linear and the marginal cost constant, whereas total cost at the coat department rose at a diminishing rate with output, yielding a declining marginal cost. In none of these cases, therefore, was there any evidence of the theoretically famous U-shaped unit cost curve. As Dean concludes (1976, p. 5):

> The rationale for the short-run cost behavior hypothesis of conventional theory is incompatible with the empirical discoveries of managerial economics in several important respects. The concept of 'the fixed factor' must be redefined both more comprehensively and more adaptively than is implied by theory. Because of several kinds of plant segmentation, the presumed invariableness and indivisibility of the fixed factor are simply inappropriate for much of modern industrial technology.

To examine the effect of scale, or size of unit, Dean studied two chains, one consisting of well over 100 personal finance offices of varying sizes and the other of over fifty shoe stores, again of varying sizes. In both cases he found clear statistical evidence of economies of scale, namely that a larger outlet was likely to reveal a lower cost per unit of output than at a smaller outlet. These conclusions seem to be supported by many other empirical studies in industry. For example, a study of Japanese ethylene plants (Lau and Tamura, 1972) suggested that while capital requirements per unit of output sharply declined with the size of plant, the requirements for labour were independent of plant size, and energy and raw materials were linearly proportional to output.

Dean discusses his results at some length (1976, p. 299), and proceeds to list ten possible criticisms to his methodology and findings (1976, pp. 303–11):

1 The findings of statistical cost–size functions are incompatible with the long-accepted theoretical model of the behaviour of firms under perfect competition . . . In the short run, costs must ultimately rise with output in order for the firm to have a determinate output. In the long run, costs must rise with size in order to preserve perfect competition, [since a continuously declining cost curve] fails to set a cost-behaviour limit to firm size, [the logical consequence of which would eventually lead to a monopoly].
2 Statistical cost findings are contradicted by facts observed in the marketplace: plants (and firms) that differ greatly in size survive despite the empirical evidence of scale economies.
3 The observed declining phase of the long-run average unit cost curve is caused by survival of the fittest; it occurs because efficient firms grow big, not because big ones are efficient.
4 The findings of statistical cost studies overstate economies of scale because firms (and plants) that are smallest, as measured by output, are likely on average to be operating at usually low rates of capacity; conversely those with greatest output are most likely to be running at full blast.
5 Accounting data on costs of firms (and plants) of different sizes do not apply valid information for measuring scale economies; omissions and non-market valuations embodied in the accounts distort true economic costs.
6 Statistical cost-size studies are expensive and of limited durability because their findings quickly become obsolete.
7 The size measures used in the studies was not satisfactory for testing the hypothesis. Mismeasurement may have distorted the statistical estimation of the relation of cost to plant size.
8 The cost effect of low rates of utilization of capacity of branch units causes the statistical cost curve that is fitted to the cost observations to overstate the unit cost of the idealized envelope curve at all plant sizes.
9 The observed size range of branch units in the two studies was narrow and the size-category of large units was thinly populated, hence the findings are unreliable for the largest store or loan office size.
10 Differences among plants (stores or loan offices) as to technology and

quality of management were incompletely removed in the studies, creating a bias toward exaggeration of economies of scale.

All these objections apply more potently to some studies than to others. There is no denying that many statistical cost studies do not provide a basis for comparing like with like, and that some of the conditions alluded to earlier are not adequately met. What is equally clear, however, is that the empirical evidence does not tend to substantiate the U-shaped cost curve postulated so widely in the literature, and that a linear cost function is often a reasonable model of short-term cost behaviour.

Effect of output on unit cost

We now turn to the question of whether general statements can be made about the effect of volume on unit cost. Suppose that a plant producing an output V has a fixed cost element of F and variable cost S, the total cost being $C = F + S$, where the cost function is assumed to be linear. The operation of the plant is characterized by what is called the 'fixed cost proportion', which is denoted by f and defined as the ratio of the fixed cost element to the total cost, namely:

$f = F/C$

For example, if $F = 1000$ and $C = 5000$, then $f = 0.25$, i.e. 25 per cent of the costs are fixed and 75 per cent are variable. The fixed cost proportion f, which theoretically can assume a value between 0 and 1, is a good measure of how capital-intensive the plant is. A high value of f suggests that the plant is highly capital-intensive, whereas a low value of f is an indication of a labour-intensive plant. Clearly, this cost proportion has a bearing on the effect of an increase in volume on unit cost (Eilon, 1984).

This effect is shown in Table 4.4 for several values of the fixed cost proportion f (the value of f is ascertained before a change in volume takes place and we assume here that the fixed cost and the direct costs per unit remain constant, irrespective of the level of output). What is interesting about these results is that they are quite general, since they are not

Table 4.4 *Effect of volume increase on unit cost (for a linear cost function)*

Reduction in unit cost (%)	Volume increase (%)					
	10	20	30	40	50	100
$f = 0.25$	2.3	4.2	5.8	4.1	8.3	12.5
0.50	4.5	8.3	11.5	14.3	16.7	25.0
0.75	6.8	12.5	14.3	21.4	25.0	34.5

confined to any particular industry or any particular size of plant. The only parameters that matter are f and the percentage rise in volume. The incentive to increase output is quite obvious, particularly for capital-intensive plants, where an increase in volume provides an opportunity to spread the fixed costs more widely. For a plant with $f = 0.75$, the effect of doubling output is a massive reduction in unit cost by 34.5 per cent, but even for less capital-intensive plants, an increase in volume has a significant effect.

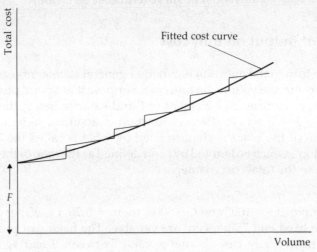

Figure 4.10 *Increments in fixed cost incorporated in the variable cost element by a fitted cost curve*

It may be argued that the assumption that fixed costs do not change is valid only for relatively small changes in volume, but that as volume increases by an appreciable amount, fixed costs would tend to rise. As the plant reaches a certain size, managerial and specialist staff increase in number, while key managers would tend to get salary increases to reflect their greater responsibilities. Analysis of remuneration of top executives in various firms suggests that there is a correlation with the size of firm (measured, for example, by its turnover). The fixed costs then increase as a step-wise function, an example of which is shown in Figure 4.10. The exact form of the increase in fixed costs has not been universally established, but for practical purposes it may be reasonable to assume that the long term fixed cost increases as a continuous function, as shown by the curve in Figure 4.10. In the case of warehouses, for example, it has been suggested (Baumol and Wolfe, 1958) that the fixed cost would be related to the square root of the throughput, i.e.

$$F = a\sqrt{V}$$

where a is a constant and V the throughput of the warehouse. If a similar relationship were to apply to a manufacturing plant, the effect of volume increase on reducing unit cost is shown in Table 4.5. Clearly, the effect is less dramatic than that shown in Table 4.4, but remains significant nonetheless. If the fixed cost increases with volume at a lower rate than that stipulated, or if variable costs per unit decline with volume (resulting from the beneficial effects of the learning curve, or through the use of more efficient production and control methods), then the expected reduction in unit costs would be greater than that suggested in Table 4.5.

Table 4.5 *Effect of volume increase on unit cost (when fixed costs increase with the square root of volume)*

Reduction in unit cost (%)	Volume increase (%)					
	10	20	30	40	50	100
$f = 0.25$	1.2	2.2	3.1	3.9	4.6	4.3
0.50	2.3	4.4	6.1	4.7	9.2	14.6
0.75	3.5	6.5	9.2	11.6	13.8	22.0

Conclusion

The main attraction of small business stems from three arguments:

1 *The economic argument* The theoretical U-shaped curve for unit cost suggests that beyond a certain point an increase in output would result in an increase in unit cost. Also, the very fact that businesses of varying sizes exist in the marketplace (the very large majority being small) is proof that small businesses are viable and able to survive against the competitive pressures of the market.
2 *The sociological argument* A small business constitutes a relatively small community, where individuals know each other and where there is an opportunity to create a congenial atmosphere that allows good communications and fosters a sense of belonging. Consequently, individuals tend to feel committed to the organization, to its products, to their colleagues and to their customers.
3 *The managerial control argument* There is no need to construct elaborate and bureaucratic managerial control procedures in a small organization. Consequently, information flow is effective and managerial decisions can be simple and quick to execute.

Leaving aside the last two arguments, which obviously have merit, there is little doubt that the economic consequences of the U-shaped unit cost curve seem to have been widely accepted as a convincing argument in favour of small and middle-sized companies. However, empirical evidence does not seem to support the U-shape hypothesis as a universal model and numerous studies, in many industries, suggest that the unit cost continues to decline as output volume increases, namely that the phenomenon of 'economies of scale' is widely prevalent. In short, it might be said that, in economic terms, empirical evidence generally points to the conclusion that 'big is necessary', particularly when the cost of innovation and product launch has to be accounted for, as discussed in Chapter 3.

This is not to say that small businesses have no long-term future in the national economy. Their very existence, even in highly industrialized countries, is a testimony to the important role that they play, and will continue to do so in the future. But as indicated by the results in Tables 4.4 and 4.5, which cover a wide range of circumstances, increases in volume can lead to substantial unit cost reductions. Against the background of fierce market competition and the need to achieve a critical mass it is no wonder that management often opts for volume expansion and increased market share as a means of consolidating or improving the competitive position of the enterprise.

Although much of what is discussed in this chapter relates to the effect of size on the unit operating cost, the basic arguments remain valid when we consider the need to marshal large financial resources to support innovation on a macro scale. As indicated in earlier chapters, the cost of innovation is only a small proportion of the total cost involved in a new product launch, which includes the preparation of a distribution network, promotion and marketing, and the setting up of the production facilities for the new product. All these up-front costs, depreciated over the life span of the product, are reflected in the fixed cost element of the total cost function and need to be recovered through the sales volume. Thus, it is through size that an enterprise can develop a capability to engage in a significant innovation activity to explore new designs, to improve manufacturing methods and to develop a new range of products. There are, of course, many other advantages of size, such as:

- Ability to develop global market strength through presence in many countries to meet existing and potential demand.
- Ability to have duplicate manufacturing facilities in close proximity to large markets in order to cut transportation costs and improve customer and after-sales service.
- Ability to source components across national boundaries and take advantage of the purchasing power that a large corporation naturally acquires in the marketplace.

- Ability to mobilize capital for re-equipping plants and for large-scale modernization programmes.
- Ability to provide varied experience and management development for ambitious young executives who look for attractive career prospects and managerial responsibilities.

The main disadvantage of size, as mentioned earlier, is the ability of management to keep control of a large and complex organization and to guard against catastrophes. Efficient communications, simplicity of decision-making processes, entrepreneurship, dynamism and flexibility – all these are said to be the hallmarks of small enterprises. The challenge to multinationals is to ensure that they too enjoy these attributes and to evolve organizational structures in which they can be fruitfully developed. We shall return to some of these questions in the following chapters of this book.

References

Baumol, W. J. and Wolfe, P. (1958), 'A Warehouse Location Problem', *Ops Res*, vol. 6, pp. 252–63.

Betts, G. G. (1982), 'Implications of Plant Scale in the Chemical Industry with Particular Reference to Ethylene Plant's, in Buzacott, J. A. et al, *Scale in Production Systems*, IIASA Proc Series, Pergamon Press.

Buzacott, J. A. Cantley, M. E., Glagolev, V. and Tomlinson R. C. (eds) (1982), *Scale in Production Systems*, IIASA Proc Series, Pergamon Press.

Dathe J. M. (1982), 'Problems of Scale in International Air Transportation', in Buzacott, J. A. et al, *Scale in Production Systems*, IIASA Proc Series, Pergamon Press.

Dean J. (1976), *Statistical Cost Estimation*, Indiana University Press.

Edwards, R. S. and Townsend, H. (1959), *Business Enterprise*, Macmillan.

Eilon S. (1979a) *Aspects of Management*, Pergamon Press, pp. 89–94.

Eilon S. (1979b), *Management Assertions and Aversions*, Pergamon Press, pp. 114–20.

Eilon S. (1984), *The Art of Reckoning – Analysis of Performance Criteria*, Academic Press.

Gold B. (1975), *Technological Change: Economics, Management and Environment*, Pergamon Press.

Gold B. (1982), 'Revising Prevailing Approaches to Evaluating Scale Economies in Industry', in Buzacott, J. A. et al, *Scale in Production Systems*, IIASA Proc Series, Pergamon Press.

Lau, L. J. and Tamura, S. (1972), 'Economies of Scale, Technical Progress, and the Nonhomothetic Leontiff Production Function – An Application to the Japanese Petrochemical Processing Industry', *Journal of Political Economy*, vol. 80, pp. 1168–87.

Pappas, J. L. and Brigham, E. F. (1979), *Managerial Economics*, Dryden Press.

Rosegger, G. (1975), 'On "Optimal" Technology and Scale in Industrialization – Steel Making', *Omega*, vol. 3, pp. 23–38.

Sampson A. (1973), *The Sovereign State – the Secret History of ITT*, Hodder and Stoughton.

Schumacher, E. F. (1973), *Small is Beautiful*, Blond and Briggs.

5 The corporate challenge

- Twelve constant threats
- The options

Twelve constant threats

As highlighted in previous chapters, the fact that a company has achieved a reasonable size in the national or international league in its industry does not mean that it is assured of survival for ever more and that it becomes immune to external threats. Changes in the economic and political environment, both at the national and international level, coupled with inevitable changes in trading conditions, in perceived needs of consumers, in working practices, in opportunities offered by new technology – all these cause a constant realignment of the competitive forces in the market. Even an internationally acknowledged leader in any given field may discover that what seems at some stage to be an unassailable position in the market may soon become vulnerable as circumstances change. Only the complacent fail to realize that any company, irrespective of size or power, is constantly under threat, which may emanate from a variety of sources, twelve of which are briefly discussed below.

Cost of materials

Sharp increases in prices of raw materials, components and energy may have serious consequences for the cost of production, which eventually must lead to a rise in the price charged for the product and consequently to a fall in demand. Some increases in the cost of materials are unpredictable and come about in the wake of political upheavals (for example, the oil crisis in the early 1970s sent the price rocketing and led to unforetold disruptions in many industries) or the incidence of wars (to which the price of metals, such as copper and chromium, and many commodities tends to be sensitive); others depend on the weather (as in the case of wheat, coffee and cocoa). The cost of imported materials is also sensitive to fluctuations in the rate of currency exchange, as discussed later. If competitors have access to cheaper sources of supply, or if they are protected from price rises by long-term supply contracts, then they gain a distinct competitive advantage.

Cost of labour

Even highly-automated companies may be adversely affected by a sudden increase in the cost of labour, which may come about as a result of:

- Increases in wage rates for regular time and overtime.
- Reduction in the length of the working week compared with companies operating abroad (this has been a major cause for dispute between the employers and the engineering unions in the UK for many years).
- Increases in paid holidays and sick leave.
- Improvements in perks and various social and other benefits.

All these, obviously desirable from the employees' point of view, lead to higher labour costs, and therefore the total cost per unit of output increases. Unless the competitors are subjected to the same pressures of increased labour costs, the result – in terms of market share and profitability – may well be detrimental. Consequently, national governments continue to exhort trade unions to moderate their wage demands, in order to ensure that increased production costs do not lead to loss of jobs. The debate on the merits and disadvantages of the Social Charter in the European Community (leaving aside the political arguments about loss of national sovereignty and its implications for the process of integration in Europe) centres around the issue of differences in employee benefits and conditions of employment in various countries in Europe, with significant consequences for total labour costs. This is why the EC Commissioners are anxious to create so-called 'level playing fields', with the same rules and social conditions for the employment of labour prevailing throughout the Community.

Productivity differentials

High wages and liberal social benefits are not the only cause of high labour costs. The other important element in the equation is that of labour productivity. West Germany and Sweden have had high wages and social benefits for many years, but they continue to retain a strong competitive position because of high levels of productivity. The productivity threat stems from the possibility that a competitor could find ways of improving his labour productivity by a quantum leap, for example by investing in new plant and machinery or in new production methods, which would then have a significant effect on labour costs.

Trade restrictions

Exporting to other countries relies on stable trade regulations, both in the country of origin and in countries of destination. If governments decide to protect their indigenous industries, or to correct an adverse balance of payments situation, by imposing restrictions on imports (in the form of volume constraints, spurious bureaucratic requirements, or increased import duties), then markets for the exporting company may shrink or even collapse, with serious repercussions on operations at the home base and on profitability of the company as a whole.

Rates of exchange

If a company has a manufacturing base in country A, purchases input (e.g. raw materials and components) from country B and exports to country C, then changes in the currency exchange rates can have a profound effect on the company's cost and revenue structure. The weakening of A's currency with respect to that of B is equivalent to an increase in the cost of imported inputs, while the strengthening of A's currency in relation to C's can have a serious dampening effect on the ability to export the product to C. A good example of the latter is the fluctuating fortunes of the British Jaguar company, which is heavily dependent on the US market and has found that its ability to export its cars to the US has, over many years, been significantly affected by the Sterling–Dollar exchange rate. There have been other examples of British companies, whose profits have been completely wiped out, and some have even suffered severe losses, as a result of adverse movements in the exchange rates. The situation is further exacerbated if sourcing of materials and components, as well as funding of the company's operations, are spread among several countries, in which case fluctuations in the rates compound the uncertainty and make the company increasingly vulnerable to forces outside the control of its management.

The cost of money

Interest rates are not just the outcome of supply and demand in the money markets. More often than not, they are directed by central banks and even by government decrees as part of their efforts to control the level of domestic demand, the rate of inflation, the level of unemployment, the balance of payments, and other key determinants of the state of the national economy. A government resorting to high interest rates, as the UK government has had from time to time, can impose a serious cost burden on its industry and put it at a distinct disadvantage in relation to competitors abroad.

The tax burden

Tax is another example of how political issues, which are totally outside the control of management, can have an overriding effect on the company's wellbeing. If the tax burden in country A is considerably heavier than in country B, then a company in the former gains a distinct advantage over a similar competitor in the latter, with the result that it can afford to pay more to its employees, to pay higher dividends to its shareholders, to invest in new facilities, to allocate higher budgets for innovation and design and to expand more rapidly.

The subsidies syndrome

Some governments are more punctilious than others in observing the rules of the free market. In an attempt to support local companies and to protect or create employment, a government may decide to resort to various schemes that provide overt and covert encouragement to local industry, such as:

- Financial support to high unemployment or underdeveloped regions.
- Low-interest loans for company start-ups.
- Free or nominal rent for plant premises for several years.
- State and/or local tax concessions for several years.

- 'Free-zone' regime, which suspends numerous rules and bureaucratic controls, including those pertaining to planning regulations and conditions of employment.
- Financial assistance in the form of long-term cheap loans for the purchase of capital equipment and direct grants to help defray the cost of design and R&D.
- Provision of government contracts that secure for the company a level of core business to help it operate near or beyond its break-even point.
- Provision of special incentives for exports (by way of cheap loans or favourable exchange rates).
- Relaxation of import controls and/or duty on imported materials and components.
- Reduced prices on inputs provided by the state (or by state-controlled agencies), such as fuel and transport.

These ten measures are common manifestations of state intervention, and they all constitute, in one way or another, a direct or indirect subsidy for a company that meets the appropriate criteria and allow it to enjoy an advantage over another company that has to operate under less favourable conditions.

Effect of microinnovation

A competitor may come up with new designs or a new product range, that could have an impact on the market. Firms in the car and computer industries are engaged in a constant battle for supremacy and regularly bring out new models, some with only marginal or cosmetic improvements over existing products, in an attempt to keep ahead of the game, or at least in order not to lose too much ground. The potential effect of microinnovation is particularly evident at an advanced stage of a typical product cycle, shown by the sales revenue for a given product in Figure 5.1, which follows from the cash flow profile discussed in Chapter 1 (see Figure 1.1). There are generally four ages in the product life cycle (often referred to in the literature as the 'S curve'):

1 The first is the age of infancy. Following the product launch, there is a 'running-in' period, during which the product is transferred from a prototype phase into full-scale production, initially on a limited scale in conjunction with market trials, and then into volume production; demand rises slowly, but prospects are sufficiently bright to encourage investment in manufacturing facilities and, where appropriate (depending on the type of product and market demand), special plants for mass production are constructed to cater for an expected large demand.

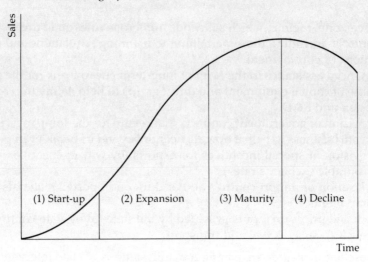

Figure 5.1 *Four ages in the product life cycle*

2 The second age is that of expansion in a rapidly increasing market, with production capacity being strained to meet the demand, so that further facilities are often installed to increase output; during this period, modifications to product design and manufacturing methods are introduced to take account of feedback from the market and to keep ahead of the competition.

3 The third age is that of maturity, when the increase in the rate of demand begins to level off, either because most of the demand for the product has already been satisfied, or because the novelty of the product has worn off, or because of new competing products on the market. Sales in a mature market increasingly depend on replacement, when previously sold products reach the end of their natural life. This is the time to convince the market that old models have been superseded by design upgrades and attractive facelifts, so that sales can be stimulated by a new demand upsurge, as shown in Figure 5.2. This has the effect of not only increasing the peak sales, but also of prolonging the age of maturity with a high level of desirable revenue.

4 The fourth age comes when sales begin to decline, partly because of market saturation and partly because of increasing competition and the effect of product substitution. This is when consideration needs to be given to replacing the old product with an entirely new model and phasing in a new revenue stream, as shown in Figure 5.3. The timing of such action, coupled with a sound policy of price differentials between the old product range and the new, depends on the rate of decline in sales of the old product and on perceived strategies of the competitors. In some cases, decline is preceded by a rather long plateau, so that the

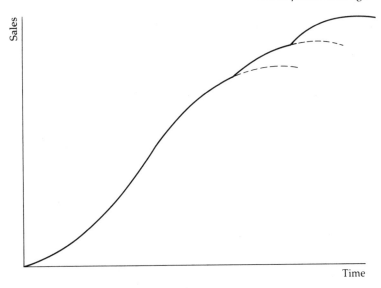

Figure 5.2 *A sales-stimulated life cycle*

product continues to be a healthy generator of cash for quite a while and a premature introduction of a direct substitute may well be inadvisable.

The pattern of the product cycle, and the time span of each of its stages, often dictate the desirable schedules for R&TD and design activities, so that the company can maintain or gain a competitive advantage in the market.

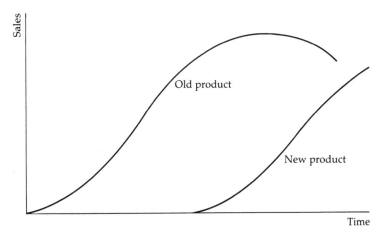

Figure 5.3 *Phasing in a new product*

Effect of macroinnovation

New technology, new materials and revolutionary new designs may render an existing product totally obsolete. As referred to in earlier chapters, the history of technology is littered with examples of products and devices that seemed to guarantee the security and prosperity of companies engaged in their manufacture, only to be faced with challenges that eventually led to serious decline and even extinction. Of the many examples that can be cited, the following few cases would suffice to illustrate this point: the first concerns developments in computer technology, which initially was concentrated on hardware and subsequently on systems development. The first generation of computers was dominated by vacuum tubes, which were too large and generated a great deal of heat; they were plagued by unstable circuits, had limited memories and their operation was slow. Consequently, it was of prime importance to achieve a high level of efficiency in programming, thereby necessitating the use of high-grade personnel. The tube reigned supreme in the 1950s but was overtaken by the transistor, which in turn was replaced by the single silicon chip. The early IBM 360 computer had dozens of switches, but by the 1970s the chip was designed as LSI (large-scale integrated) units which could accommodate hundreds of circuits, only to give way in the late 1970s to complex VLSI (very large-scale integrated) designs with thousands of circuits and later to multitask, multichip architecture, so that now designers talk about ULSI, SLSI and FLSI configurations (ultra, super and fabulous). It is estimated that in the 1950s and 1960s the average improvement in CPU (the central processing unit of a computer) and memory amounted to 80 per cent per annum coupled with a reduction in cost (for a given performance) of 55 per cent (Friedman, 1989). Another innovation example, on a more mundane level, is the demise of the slide rule, which first suffered from progress made by electromechanical desk calculators and then was rendered a mortal blow from electronic pocket calculators. A third example is the story of the steam locomotive in the US and in the UK (Hawthorne, 1978), shown in Figure 5.4. The number of steam locomotives in use rose steadily, and at times rapidly, over a period of 100 years, before the challenge of diesel laid steam to rest. In aviation the propeller engine was supreme until the advent of the jet engine, which, in spite of early scepticism, took over completely, at least for medium and long-range hauls. In manufacturing, plastics have replaced metals in countless applications and have opened new exciting opportunities for novel designs. Against this brief chronicle of change, no company can regard any of its products or designs as safe from obsolescence and extinction.

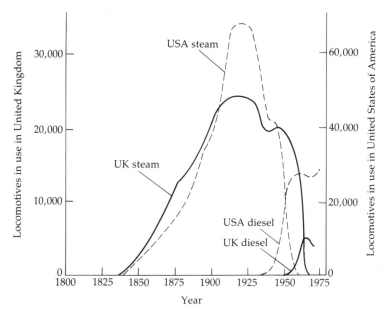

Figure 5.4 *The number of steam and diesel locomotives in service in the UK and the USA*
Source: Hawthorne, 1978

Distribution and service

The impact of innovation is not confined to design and manufacturing, but to distribution channels, packaging and customer service (pre-sale and post-sale). If a competitor suddenly announces the introduction of novel distribution and service methods, the effect on the market may well be significant. In some industries, promotion and distribution are distinctly more important than the intrinsic characteristics of the products offered for sale.

Takeover

There is always a danger that someone is plotting a takeover. In the past, big companies were thought to be immune from the threat of takeover, simply because of the vast funds that would need to be mobilized to make a bid. This has been at least one of the reasons for companies striving to become bigger and for the intermittent popularity of conglomerates since the Second World War. In recent years, however, mega takeovers involving billions of dollars or pounds have become commonplace. Formation of ad hoc syndicates, coupled with the intricate design of suitable financial instruments (heralded by the junk bonds in the US), has made mega bids possible. The takeover of the Macmillan Publishing Co in the US by Maxwell Communications and that of Consolidated Goldfields in Britain by Hanson Trust are just two examples of successful bids that were considered impossible even a short while before they materialized. Many such bids are followed by a move for rationalization and a wave of asset stripping to help finance the bids and to yield quick returns to the successful bidders. The threat of a takeover is, therefore, a constant reminder to the senior executives of an organization of any size that their own position, power and privileges could all be at risk.

In the light of these constant threats emanating from all quarters, many companies fully realize that the economic environment is not static, that because change is inevitable, coping with it is a skill and attitude of mind that must be inculcated throughout the managerial hierarchy. Some companies can adjust more readily and more swiftly to certain changes than others. Marketing-oriented companies, particularly attuned to customer behaviour, are geared to detect market trends and adapt their promotion and pricing strategies accordingly, while production-oriented companies can take advantage of changes in exchange rates and prices of inputs by switching sources of materials and production facilities from country to country and thereby remaining cost competitive. As a result, companies constantly gain and lose competitive advantage, depending on the kind of changes that take place in the economic environment and on the ability of these companies to cope with change.

The options

What can a company do to mitigate the threats enumerated above? The answer is either to create conditions which reduce the probability of a threat materializing, or to convert a threat into an opportunity and turn the tables on the opposition. The main strategic alternatives are the bootstrap option, the acquisition option, the merger option, and strategic alliances, and these options are discussed below.

1 The bootstrap option

First, the management needs to consider how the company's own resources can be used to improve its competitive position, through:

- Capital investment aimed at introducing new production methods, or improving existing methods, thereby reducing the labour content of the product.
- Promotion campaigns aimed at increasing market share and volume, which will in turn help to reduce unit costs and make the product potentially more profitable.
- Improvement of existing distribution channels and liaison with customers, to ensure the provision of effective pre-sale and post-sale service and advice.
- Concentration on product quality and reliability to increase customer awareness of the company and its products, and thereby achieve greater market penetration.
- Allocation of resources to design, which is a multifaceted activity, as shown in Figure 5.5, to meet the operational expectations of the user (including the aesthetic characteristics, which may well prove to be crucial), and to improve the functionality of the product, its durability and dependability, its flexibility and its ease of maintenance, while at the same time taking account of production constraints and opportunities (Eilon, 1962).
- Macroinnovation with the mission of yielding a new generation of products to completely replace the current range, and steal a march on the competition.

Needless to say, these approaches are not mutually exclusive. For example, capital investment in production facilities is often combined with measures aimed at increasing volume, with the dual purpose of justifying the investment and reducing unit cost. Similarly, an integrated strategy for investment in distribution methods, quality, design and R&TD can have both defensive and offensive characteristics in relation to the competitors. Needless to say, all these measures demand considerable financial resources, as well as high calibre personnel, and above all – time. Few companies enjoy the luxury of being able to afford all three, in which case they have to consider some of the other available options.

2 The acquisition option

When time is a serious constraint to the speed with which a company can institute internal changes, or raise additional financial resources in order to address the issues listed above, then a takeover of another company may well be the answer. Consider the case of a company concluding that it must strengthen its distribution network, or that it needs to penetrate a new market. To achieve its aim through organic growth is often a very costly exercise; furthermore, it may take years to accomplish. Identifying a company that already has a well-developed network of outlets in the target market, and making a takeover bid for that company, is often a much more effective route.

As examples we note the takeovers of several bank networks in the US

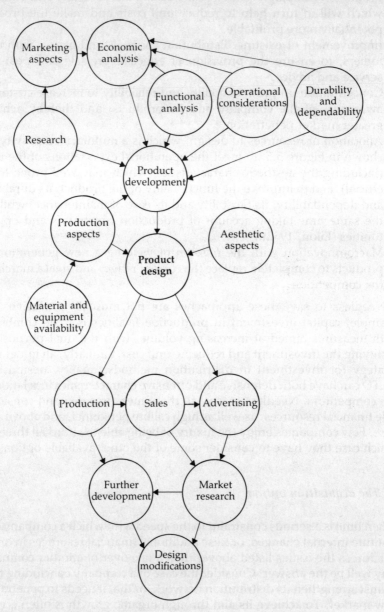

Figure 5.5 *Some interrelations involved in product design*
Source: Eilon, 1962

by British and Japanese banks, the takeovers of many family firms in the building materials industry to create large companies with greater muscle in the marketplace (such as ARC and Tarmac in the UK), the takeovers of high street retail chains to create large networks of outlets, and, in the same vein, the takeovers of hotel and restaurant chains. All these takeover and merger moves are equivalent to acquisition of a market share (in an existing or in a new market), they provide opportunities for rationalization and economies of scale, and they underline the potential of improved profitability.

In the manufacturing area, the acquisition of another company may not only buy its market share, but also provide modern manufacturing facilities nearer to the point of sale. Advanced manufacturing methods, new product designs, new technological developments, patents and related know-how – all these become instantly available when a company that possesses such attributes and assets is acquired.

Another important consideration of company takeover, and just as important as the physical assets and market share, is the acquisition of technical and managerial manpower. Growing a cadre of highly-skilled staff, able to face new challenges, can be even more costly and time-consuming than creating new physical facilities. If the target company has good technical staff in the areas of manufacturing, design, R&TD, or marketing, or if it has high calibre managers who could potentially take on wider responsibilities in a larger enterprise, then the acquisition could well be motivated by these considerations alone.

It should be appreciated that one aspect of a company acquisition (when the victim company operates in the same area as the predator) is that it eliminates a competitor. This is why an acquisition may be regarded as both a defensive and an offensive act. It is defensive in attempting to reduce the potential threats enumerated earlier, and it is offensive in trying to eliminate the competition altogether.

3 The merger option

If a suitable acquisition is not available, a merger may be considered. The difference between a merger and an acquisition is that in the latter case the predator determines the shape and composition of the enlarged company, its policy and strategy, and the fate of all the executives of the victim company. A merger is usually a marriage between equal, or near equal, partners; there need be no victorious predators and loser victims. The purpose of the merger is to create a new entity and to exploit its potential synergy, so that its combined strength is greater than the sum of the separate parts. A good example of that is the merger in the early 1970s of the National Provincial Bank and Westminster Bank to form the National Westminster Bank, which has become one of the top banks in Europe, with extensive operations in the US and elsewhere. Many other examples were mentioned in Chapter 3.

Clearly, a merger can only succeed if the partners are compatible in their respective cultures and visions. Even then, there is the prospect that not all the individual aspirations of the senior executives can be met. Admittedly, the merged company would operate on a larger scale than either of its constituents, and the scope of many jobs in the new organization may be enlarged, but there would be a need for only one chairman, one chief executive, one finance director, one marketing director, and so on. Some executives would gain in status, some would not; some directors would remain on the board of the merged company, some would lose their seats. Such uncertainties rarely preoccupy the senior executives of a predator company in the case of a takeover, and perhaps this is why many would-be predators prefer takeovers, whereas many would-be victims prefer mergers.

4 Strategic alliances

The general scenario of a board considering strategic options develops as follows:

- Can we go it alone? What are our main strengths and weaknesses? What are the possible threats to the company and from what quarter?
- What can we do with our own resources to meet these threats? If we do not have the necessary resources (skilled personnel, production capacity, machinery, space, production and marketing expertise, technical know-how, effective administration and management), can we buy them? If money to buy is not immediately available, can it be raised without loss of independence? What would be the consequences of going it alone?
- If adequate financial resources are available, or if money can be raised, but if the desired facilities cannot be acquired within a reasonable timescale, can we buy a company that already has these facilities in place? If so, which companies would be a 'good fit' and should be identified for the purpose? Are they available and at what price? What would be the expected results and opportunities of a takeover and within what timescale? What are the down-side risks and how can they be avoided?
- If no suitable companies are available for a takeover, or if the potential victims are too small to satisfy the grand design for expansion, is there a formidable competitor, roughly of our size, with whom we could

merge? Would such a merger be allowed to proceed, given current anti-monopolies legislation? What would be the conditions for such a merger to succeed? If a merger does go ahead, what would be the consequences in terms of:
- loss of independence.
- the timescale of adjustment and restructuring.
- the future of key executives in the merged organization.
- possible retaliation by remaining competitors.

On reflection, big companies often find that none of these courses of action can adequately meet their future needs, particularly when they aspire to become major players in the global market. The bootstrap option may take too long to implement, or may be too risky in the light of technological advances already achieved by competitors; proposed takeovers may be too small to make a significant impact on desirable expansion plans; and large-scale mergers may be too unwieldy to manage, or they may be delayed or blocked by powerful government agencies.

The route that needs to be considered then is that of strategic alliances, some aspects of which were discussed in earlier chapters. A strategic alliance is an agreement between two or more parties to collaborate in specified areas. The agreement may have a fairly limited scope in allowing a certain exchange of information, or it may involve extensive sharing of marketing, manufacturing and laboratory facilities. It may be confined to particular products and markets, or it may be far reaching and extend to global collaboration. It may be specifically designed to handle joint ventures relating to a new range of products and development of a particular market, or it may be aimed at sharing the ever-mounting costs of macro-innovation, not only in R&TD but possibly design, production of prototypes and field testing as well. It may be formed as a nucleus of a consortium with the intention of attracting further partners in due course, or it may be set up as an exclusive alliance. The number of variations on the theme is enormous.

Alliances have their advantages and disadvantages, some of which are summarized in Figure 5.6. The main advantages are immediate access to information and technical know-how, as well as to production facilities, all of which may otherwise take a great deal of time and financial resources to acquire. In addition, alliances often lead to valuable marketing agreements (including those relating to pricing policies and the avoidance of damaging price wars) and effective coordinated promotion campaigns. The disadvantages centre around the need to share hard-gained technical expertise, to reveal secret advances in innovation, and to face the serious risk of severe constraints when managers come to consider future product development and marketing strategies. Alliances certainly help to contain and even eliminate external threats, but they also involve commitments and obligations. In short, alliances offer new opportunities and increase corporate muscle, but they inevitably curtail freedom of action.

After considering all the pros and cons, an alliance comes into being if the advantages are perceived to outweigh the disadvantages. And this perception needs to be shared by all the partners involved. An alliance is

Advantages	Disadvantages
Technological	
Access to the ally's latest: • technical information • product design specifications • R & TD results • developed technical software Sharing expenditure of R & D and design Ability to use special testing facilities	Giving away information • after heavy expenditure • costing time and money Possible misuse of the information by the ally Fear of leakage to third party Compatibility of standards and specs
Marketing and sales	
Access to new markets Benefit from ally's reputation Sharing promotion expenditure Using an ally's established • distribution network • after-sales servicing facilities Price agreements and avoidance of price wars Use of known trademarks	Giving ally access to own market Constraints on volume and degree of market penetration Strain on own distribution network to serve ally's needs Loss of control on own trademarks Constraints on promotion strategy
Production	
Information on new production processes Access to ally's production facilities Cheaper sources of components/subassemblies Access to product assembly lines Benefit from larger purchasing muscle Ability to manufacture under licence Royalties from selling manufacturing licences	Being reliant on ally to supply Constraints on the use of other suppliers Problems with quality
Financial	
Opportunities for equity participation Attractive sources of finance for joint ventures Opportunities for improved cash flow Reduced currency exchange exposure	Some loss of financial freedom Constraints on investments
Managerial	
Sharing experience Staff exchanges Sharing training and management development programmes Representation in geographical areas where own full time staff would be costly	Serious difficulties in managing joint ventures Need for elaborate mechanisms for coordination Constraints on developing own programmes

Figure 5.6 *Advantages and disadvantages of alliances*

not a zero-sum game, in the sense that what one partner gains the other (or others) must inevitably lose. An alliance makes sense if, in spite of the enormous effort that is required in planning and coordinating future activities involving many managerial functions, all the partners expect to gain from combining forces to create a competitive advantage over the other operators in the market. This is particularly the case when the alliance involves a joint venture, to which financial, physical and manpower resources need to be allocated. The parent companies are naturally reluctant to lose their best technical and managerial staff to the new venture, but without a whole-hearted commitment, it may never succeed, or its progress may be seriously delayed.

One of the most difficult aspects of alliances is that of disengagement. An alliance may work perfectly satisfactorily for a while, but with the passage of time it may face new problems: Management changes at the partner companies may dent the zeal with which the original alliance was pursued; the effectiveness of the existing alliance may be affected by new alliances set between competitors; a new alliance struck by one of the partners with a third party may threaten the profitability and even viability of an existing alliance; new governmental regulations and changes in trading conditions may affect future prospects; and so on.

When changes in the economic and competition environment cause the rationale for an existing alliance to be questioned, the partner seeking a 'divorce' may find that the process of disengagement is fraught with numerous financial and legal difficulties. The dilemma for management is that when it comes to consider an alliance in the first instance, it needs to decide on how binding it should be. A very loose relationship may have the advantage of easy disengagement, but its benefits may not be substantial. On the other hand, distinct advantages may be gained by a genuine pooling of forces and a high level of commitment on behalf of the partners, but then a very high price would have to be paid for disengagement (assuming it remains possible). As in all managerial decisions, consideration of alliances must weigh short-term prospects against long-term problems.

At the end of the day, though, there may be little choice, if other strategic options are not available, or are not likely to be effective. A decision not to proceed with strategic alliances may result in hostile alliances being formed by the competitors, with dire long-term consequences for any company left out in the cold. Alliances have thus become popular among big companies in recent years, as the case of the telecommunications industry demonstrates (Devlin and Bleakley, 1988). A similar process for forming alliances in the automobile and telecommunications industries was outlined in Chapter 3 (see Figures 3.3 and 3.4), where the rationale behind the Volvo–Renault link-up was discussed. It is interesting to note that the movement in the post-war years to create large companies in the automobile industry has not precluded the establishment of alliances in a series of defensive and offensive actions.

The intricate relationships between big companies, such as the examples shown in Figures 3.3 and 3.4, are reminiscent of political alliances between governments over many centuries, some distinctly defensive in character, others aggressive and expansionist. As discussed in earlier chapters, industrial alliances are increasingly evident in the area of macroinnovation, where huge financial and technical resources (including highly-skilled scientists and technologists) are needed, not only for the next generation of products, but for sheer survival. The battleground is tomorrow's marketplace, where the protagonists are constantly striving to gain or maintain a satisfactory market share. Without strategic alliances in the field of macroinnovation, few companies can hope to survive, and we shall return to this issue in the next two chapters.

References

Devlin, G., and Bleakley M. (1988), 'Strategic alliances – guidelines for success', *Long Range Planning*, vol. 21, No. 5, pp. 18–23

Eilon, S. (1962), *Elements of Production Planning and Control*, Macmillan.

Friedman, A. (1989), *Computer Systems Development: History Organization and Implementation*, John Wiley & Sons.

Hawthorne, E. P. (1978), *The Management of Technology*, McGraw-Hill.

6 Governments and innovation

- Intervention
- Investment in education
- Public purchasing
- Subsidized industrial research in the UK
- The UK Government's Alvey programme
- Subsidized industrial research in Japan
- Offset programmes
- Subsidized inward investment
- Subsidized exports
- State-financed companies
- Airbus Industrie
- Space programmes
- Industrial policy

Intervention

State intervention in industrial affairs waxes and wanes. At the present time it is out of fashion, particularly in the UK. Few governments in the 1980s can have campaigned more vigorously than Mrs Thatcher's against interventionism and this counter-revolution against the practices of the previous decades has spread in varying degrees around the world – even into the fastnesses of what was the Communist bloc. Nevertheless, even at this nadir of state intervention, much remains and the fundamental dilemma of economic versus social considerations is unlikely to go away. In this respect it is instructive to contrast the attitudes prevailing in 1933, in the aftermath of the Great Depression, when even the United States under President Roosevelt was adopting the interventionist measures of the New Deal, with those in 1985, when few nations were not enjoying an economic boom.

In 1933, an international conference under the auspices of the League of Nations met in the Royal Institute of International Affairs in London. One section of this was devoted to 'State Intervention in Private Economic Enterprise' (League of Nations, 1933). The case, forcefully and logically put, was advanced that liberalism (as they then described free-market economics) had produced unacceptable social problems and that its inherent determinism (i.e. that governments were powerless in the face of market forces) was an affront to the human spirit.

In 1985, David Henderson gave the BBC's Reith Lectures (Henderson, 1986). (The BBC in 1933 was held up as a model of a state rather than private sector enterprise!) From his background as a professional economist in the UK Treasury and as the head of OECD's economics and statistics department, his theme was a powerful defence of classical economic liberalism and an attack on political interference in the free operation of market forces. The latter he characterized as being based on 'Do-it-yourself economics' (DIYE) and claimed that its influence extends across the whole political spectrum. Nevertheless, even he conceded that the pressure on politicians to intervene is understandable – and irresistible – and that the most that he, as an economist, could do, was to point out the more disastrous consequences that certain forms of intervention would have.

Intervention takes two main forms – regulatory and supportive. *Regulatory intervention* has a powerful effect through the legislative and taxation roles of the state, monopoly control and health and safety requirements and through the quasi-independent bodies approved by the state to regulate the financial markets and company behaviour. These, as we shall see in Chapter 8, present transnational business with a confusing array of conflicting national requirements.

Supportive intervention takes many forms and has a powerful influence even in the most non-interventionist political environments. Most remote from the point of sale of innovative products, and yet of prime importance, is the educational system. Large elements of this are targeted at supplying industry with the skills and professionalism of innovation and there is much open debate on how this should best be achieved. Linked with this are the research activities of universities, polytechnics and government-maintained institutes, usually receiving their basic finance from the state but augmenting this with specific contracts from industry.

From this point there are a variety of ways in which industrial innovation may be subsidized through state research and development contracts and through the financing of innovative products through public purchasing. In the international scene there is the use of regional development funds, offset requirements and overseas aid to benefit domestic industry. Finally, there is direct financial assistance from loans for specific projects or through the provision of working capital for state-controlled enterprises.

Investment in education

However non-interventionist a government may be, it is inevitably committed to financing the educational system of the state. In so doing it has to reflect both the social and economic requirements of the community and to balance those subjects that enhance the quality of life with those that generate the wealth on which to a significant extent that quality depends. Despite strenuous endeavours to separate wealth creation from those activities which enrich life, it is now generally accepted that they are inseparable and that both march forward together.

Investment in those aspects which contribute to wealth creation is, therefore, of great importance and, of course, the value of such investment is dependent on the understanding of the wealth-creating process of those who spend the money. An objective examination of this subject is contained in a report prepared for the British Government by a Committee of Inquiry it set up under Sir Montague Finniston (Finniston Committee, 1979). This report critically examines the formation of engineers in the UK, Germany, the USA, Japan etc. While much is concerned with matters specific to the UK, it confirms that the systems used in these countries in large measure reflect the national attitude to industry and particularly to the role of innovation in it.

Japan, for example, sees engineering as the linchpin of a commercial enterprise, using innovation as the key to market share and the means of producing the product at a price and with a quality which will maximize its penetration. Germany, without the same dedication to market share, sees engineering as the dominant consideration in the company determining the company's overall strategy as well as carrying out specialist functions. The scene in France is set by the Grandes Écoles with their emphasis on mathematical and theoretical excellence coupled with preparation for top industrial and government administrative appointments.

Britain, with its long tradition of education for leadership and administration in Britain and its overseas Empire, 'emerges as a nation that is struggling to escape from the treating of engineering as a specialist application of science with its role limited to providing expert services to a technologically uneducated higher management of leaders and administrators.' This, in significant measure, explains the imbalance between the acknowledged capability of Britain to initiate innovative products – the feasibility stage of the innovative chain – but its failure to exploit them commercially, because of the lack of 'whole men' and 'total understanding' to secure the finance and the markets which turn an idea into a profitable business.

The distribution of global enterprises driven by competitive innovation must inevitably be influenced by the availability of appropriate talent in the countries involved. The national participation in such global activities will, therefore, depend on the quality and magnitude of the state's investment in the education and training of such people (more on this in the next chapter).

This puts a major element of a state's educational programme and associated investment into global competition for a share of the innovative industrial activities of the world market. It is no longer a matter of internal cultural preference, but bears directly on the degree to which the nation gets its share of transnational industry. The debate on the resources allocated to this task should, therefore, be conducted in recognition of the competition from other states in this respect, the need to 'sell' the merits of a national educational system on the global market and the economic consequences of being uncompetitive.

The considerations apply equally to the government funding of research in universities, polytechnics and other national institutions. Research and its associated term 'arandee' in relation to the technology-base of innovation is a dangerously ambiguous description of widely different activities. In Chapter 7, we discuss the definitions used by the EC in their interventionist programmes.

The OECD in their Frescati manual, have struggled to distinguish between the various different sorts of research which governments are involved in. The initial framework set down contains basic research (no application in view), applied research (directed at a practical application) and experimental development (directed at a specific application). Importantly – but unfortunately for simplicity of definition – there runs through all this the concept of strategic research. Once research – basic or applied – has to make its case for money in competition with others, the subject of an eventual economic pay-off, no matter how remote, arises and the strategic dimension is invoked. This may relate to 'market pull' ('if only a material could be found with such and such characteristics') or 'technology push' ('if only electric power could replace the internal combustion engine in the motor car').

All this has a major influence on the higher echelons of the educational system. To educate at this level it is not enough to teach the corpus of yesterday's established knowledge. Those involved must be participating in the creation of tomorrow's technology if those being educated are to leave 'up to speed' and ready to make their contribution in the innovative economy. Government 'subsidy' of this research element of higher education is vital and should be viewed as part of the competition to attract the world's innovative industries – and a major factor in deciding the national participation in them.

From the industrial point of view, however, this type of research plays a relatively small part in the competitive innovation scene. In the main it relates to published work and contributes to the store of common knowledge available to all competitors, and this must be so if the motivation that has brought so many advances in science and technology from this source is to be preserved. It is, therefore, in our view a mistake for governments to try to shift the burden of financing a significant part of university research on to industry. Exclusive access to the results is almost inevitably required by the private sector and is the antithesis of the academic method of publication and criticism by their peers.

There is, nevertheless, a need for close contacts between universities and industry, but this is best done outside the mainstream of research in what might be described as consultancy work. Here the university staff, drawing on their knowledge or access to knowledge, are brought in to advise on an industrial research programme and may well augment this with further investigations specific (and confidential) to industry's problems. Each party gains intellectually from such arrangements and the school or faculty earns money which it can use to augment that received from the state. The point is that this is a bonus and no substitute for the recognition that the state's investment in education and educationally-related academic research is directly related to its position in the global competition for innovative industry.

Public purchasing

Governments have both a direct and an indirect influence on innovative industry through their public purchasing activities. The sheer volume of their requirements represents a very important launch platform for those who are fortunate enough to secure a government contract. However, even more potent, although not always beneficial, is their willingness to subsidize some parts of the innovative chain required to produce the product required.

In the name of special national requirements and, in the case of defence, national security, the public purchasing agency takes command of a substantial part of the innovative chain. It may make a substantial contribution to the feasibility research, dominate the definition phase, determine when or how the product will be produced and control the form of after-sales service to suit its requirements. In this way, the all-important consideration of what other customers may require is suppressed, customer need is subjugated to the execution of a predetermined specification in which the enterprise has a minor role and the contractor, which in this approach is what the enterprise becomes, is solely an executer of work defined by the customer.

The damage this can do is amply illustrated in the book *Britain's Economy – the Roots of Stagnation* (Jones, 1985). The author, Aubrey Jones, held ministerial level appointments in successive governments in the UK in the 1950s and 1960s and presents, wittingly or unwittingly, a catalogue of the failure of such public-funded enterprises. Civil aircraft designed and manufactured to a specific UK requirement, nuclear power development ending in the demise of the Advanced Gas Cooled Reactor and, of course, Concorde, are just a few of the examples he provides of the failings of public procurement in the generation of commercially successful innovation. In the chapter headed 'The view from within British industry' he opens with the statement: 'It has been demonstrated in the chapters on energy and aerospace that the private contractors to government were careless of costs and inept at marketing.' Is this altogether surprising in an essentially cost-plus relationship with a single customer who, because he was putting up the money, took upon himself many of the key roles normally played by the top management of a company?

In defence, where the problem is both most intractable and acute, all governments have been aware of this problem and have sought a variety of solutions over the years. All have had very limited success because the premise of competition in its full sense could not be accepted. The market is war and the idea of companies, particularly outside the boundaries of a nation, taking the initiative and proposing more effective ways of conducting war – defining the product, marketing it worldwide and being rewarded for successful initiatives with large profits – is political dynamite.

Nevertheless, the UK Government, in its policy statement *Value for Money in Defence* (MOD, 1983) seeks to introduce some of the benefits of competitive innovation into defence procurement and this concept has now been accepted by the so-called Independent European Programme Group (IEPG) of the European Defence Ministers of the NATO alliance.

The first prerequisite of any semblance of competition is that there should be a plurality of competitors. With the highly-specialized nature of defence equipment manufacturing, this is difficult to achieve even in the USA, and nearly impossible in countries like France, the UK, Germany and Italy and out of the question in the rest of the Western alliance. It can only be achieved by transnational companies and joint ventures.

The second prerequisite is a significant degree of commonality in the perceived needs of the member nations. This can be realized by segregating the vital specialities which characterize, and indeed justify, the independence of thinking of national forces from the less vital commonalities in which industry can claim superior knowledge. In the case of an aircraft or missile-based weapon system, the 'platform' as it is called – the carrier – can be in the second category, whereas the specific method of finding and destroying the target may well be a secret speciality of a particular nation. The 'platform' is then the element potentially capable of competitive purchasing. This subdivision accords with experience both in that 'platforms' continue in service with a variety of updated mission equipment and in exporting to nations without a defence industry where this variety often obtains.

The third prerequisite is that the competing enterprises must develop, from a strong market research and intelligence base, their own technological research and feasibility demonstration programmes, largely if not entirely financed on a speculative basis. Only in this way can they hope to enter the competition with a proven technological, cost and delivery date edge over their competitors. Sufficient understanding of all possible markets inside and outside the alliance must be assessed in making such speculative investments. Furthermore, it must be anticipated that not all competitions will be won and so a sufficiently wide spread of outlets is needed to get as far away as possible from the dangers of a 'one product' company, referred to in Chapter 2. It will be clearly beneficial if some of these can utilize the technology base derived from defence work.

In short, this competitive defence company of the future must be a 'full-blooded' multinational enterprise. The same considerations apply in the other areas of public procurement if the benefits of competitive purchasing are to be achieved – a declared goal of the European Community.

There is, however, another important consequence. Much has been made of the disadvantages accruing to those nations who are, traditionally, the innovators in defence equipment. The technologies concerned, acquired at great expense in taxpayers' money and diversion of the nation's stock of technological talent, have a disappointing record in terms of commercialization in the civil market. The enterprises required to respond to truly competitive purchasing of defence equipment can be expected to be driven to become much more capable of exploiting the technology base than the specialist nation-based defence contractors of the past. Investing their own resources in anticipatory research generates the pressure to make it as widely usable as possible. Similarly, innovation in production, rarely present in an old-style defence contract, becomes a necessity if cost and delivery are accorded competitive importance.

The pressure for efficiency in public procurement, the growth of transnationalism in the industrial world and the realization of the importance of massive public procurement in the evolution of innovation all point to a significant movement away from subsidized R&D in the future. As it develops, the enterprises that compete in this market will inevitably be big, multinational and quite different from the national contractors to governments of yesteryear. With this change, this particular brand of government-funded and managed R&D will decline to a minor, specialist role in the global innovative scene.

Subsidized industrial research in the UK

Quite separate from the funds invested by government in the procurement of products for their own requirements, there is a long tradition of government funding of research directed at industrial products in general. In aviation, in agriculture, in materials, etc., a combination of government institutions and the research departments of large industrial concerns has received taxpayers' money to this end. As the acid test of commercial benefit is increasingly applied to justify such subsidies, an interesting variety of national policies is emerging.

The UK Government, through its Department of Trade and Industry (DTI) and in accordance with its anti-intervention policy, represents an extreme position of a state in this respect. In a paper delivered in Brussels in 1989 by Dr Coleman, its Chief Engineer and Scientist, this policy was stated in the following terms (Coleman, 1989):

> Innovation is the key to profitable sustainable growth. R&D is an important component of innovation but R&D alone is not sufficient for innovation to occur. We must not forget the importance of other parts of the innovation process if R&D expenditure is to be exploited effectively and profitably.
>
> The overall aim of the DTI's innovation policy is to encourage a net addition to innovation by industry without creating or perpetuating distortions in the economy. I see no reason to expect this policy to change post-1992.
>
> Individual companies will continue to be the principal source of R&D funds mainly from retained profits and this is particularly so for near market activities. Last year my Department reviewed its role in encouraging innovation. The main conclusion was that the balance of existing policies should change in order to move away from near market R&D support.
>
> Firms themselves are best able to assess their own markets and balance the commercial risks and rewards of financing R&D and innovation. The closer to the marketplace that innovation is taking place, the more fundamental this should be as a guiding principle of policy. The DTI's innovation policy should be focused on circumstances where research is necessary before commercial applications can be developed or where the beneficiaries of the research are likely to be widespread.
>
> The only exceptions in the UK are for projects which are so large, relative to company size, that the company's survival depends on the outcome of the R&D. Companies find it difficult to raise adequate finance for such high-risk activities. Hence the continuation of single company support for aero-engine and airframe manufacturers and launch aid for major aeronautical projects. At the other extreme, very small companies may be unable to raise modest sums of capital for innovative projects because of the relatively high appraisal costs compared with capital required and the return on capital expected for high-risk projects.
>
> Over the next decade the need for public support for the European aircraft industry may reduce through stronger strategic business alliances between companies. For small companies some element of risk sharing is likely to

remain but as financial institutions gain greater experience of new high-tech companies in Europe, I expect the private sector to accept the role of banker more readily as already happens in the US.

The key role for governments and inter-governmental agencies is in developing the right climate to encourage innovation in business. Post-1992, the role of the EC will increase because of the harmonization of regulations and standards.

This is wholly consistent with the transfer of responsibility for the whole innovative chain to industry and with regarding a government subsidy of the R&D element as the exception rather than the norm that prevailed not only in the UK in the post Second World War decades up to 1970.

The definition of what is exceptional, however, is a matter of judgement and it will not be the same in every country. Most importantly it is a judgement made by governments and, therefore, probably with political goals and expediencies in mind. Thus, the escape from 'government knows best' of the interventionists is incomplete. The 'special case' of the aircraft industry is mentioned in the above quotation. The 'even more special case' of the embryonic space sector and other 'way out' activities that flow from it is another example. To these big special cases can be added the mass of relatively small 'pre-competitive' research activities partially subsidized by the state 'where research is necessary before commercial applications can be developed or where the beneficiaries of the research are likely to be widespread' (Coleman, 1989).

The UK Government's Alvey programme

It is instructive to look at the history of the 'non-interventionist' UK Government's programme in the information technology sector – the so-called Alvey programme. This was a programme designed to advance the position of British industry in the burgeoning information technology market as a special case. Launched in 1982, with a total commitment of some £350m from Government and the industrial participants for a period of five years, it was given special recognition and support through a new appointment, a Minister for Information Technology in the Department of Trade and Industry.

By 1986, with a review in the offing of what the next stage should be, Brian Oakley, the Government official in charge of the programme, set out its achievements (Alvey Directorate, 1986a). A total of 110 companies were engaged in some 300 projects. Almost every British university and eleven British polytechnics were involved, with academic partnerships with industry occurring in 85 per cent of the projects.

However, of the companies taking part, GEC was involved in fifty-nine, ICI in forty-nine, British Telecom in thirty-seven and Plessey in thirty-five. Imperial College topped the list of university involvement with thirty-four projects, followed by Edinburgh with thirty-one and Cambridge with twenty-four. Clearly, the big companies with sales that could easily bear their slice of the gross investment of some £70m per annum in the programme and a few universities who needed no introduction to such companies constituted a large part of the achievement.

A Committee – the 'IT '86' Committee – was then established under Sir Austin Bide, then Chairman of the highly innovative pharmaceutical company Glaxo, to examine progress and to make recommendations on what should be done, as the end of the initial programme was then in sight (Alvey Directorate, 1986b). In short, it recommended that a further major programme be launched, at a total cost of about £1bn, to encourage application of the technology acquired and to generate more for the benefit of both suppliers and users.

In a climate of belief that industry should be responsible for industrial R&D, it was not surprising that the Government response was a long time coming and that there were indications that this course of action would not be accepted. In 1987, there was a General Election which, while the same political party returned to power, resulted in the appointment of a new Secretary of State for Trade and Industry who openly attacked the support of industrial research, and the appointment of Minister for Information Technology was abolished. The 'IT '86' recommendations were effectively lost and such activity as remained was absorbed in the European ESPRIT programme (see Chapter 7).

The London Business School talked of it being wrongly focused (Grindley, 1987), discouraging British companies from seeking research partnerships overseas. The National Audit Office, an independent organization funded by the UK Parliament, issued a report criticizing the lack of exploitation of the funded research (National Audit Office, 1988). Among its criticisms were:

- Heavy spending on administration and support facilities.
- Long delays in bringing together teams to work on projects.
- Five large UK companies accounted for more than half the total partipation in the scheme – although a substantial amount had been achieved in terms of close cooperation between industry and academic institutions.

This report was followed by the all-party Parliamentary Committee on Science and Technology criticizing the Government for not taking a more effective role in public procurement of information technology products and for abolishing the post of Minister for Information Technology (House of Commons, 1988).

In defending its achievements, the Alvey Directorate claimed a commercial success with the launching of a new static random access memory with very fast processing speeds. The total investment involved was £6.3m, shared between the Directorate, British Aerospace, Racal and Standard Telephones and Cables (STC) who now seek applications in the telecommunication and defence sectors of information technology.

It is difficult to avoid the harsh conclusion that the 'mountain' of this programme had, in commercial terms, produced a 'mouse' of industrial innovation.

Subsidized industrial research in Japan

Government intervention, Japanese style, has been the subject of much study – and much misunderstanding. MITI, the Ministry of International Trade and Industry, has had attributed to it a whole range of interventionist practices and powers which careful study seems to reveal to be quite untrue. Daniel Okimoto in his book *Between MITI and the Market: Japanese Industrial Policy and High Technology* dispels a number of myths while usefully illuminating the special role which MITI plays in the Japanese system (Okimoto, 1989). He shows how MITI has sought to create a supportive environment for business rather than to influence the performance of companies directly. This view was reinforced at a Tokyo seminar in 1987 given by the Japan Technology Transfer Association. Mr Hajune Koratsu, a professor at Tokai University and a technical adviser to Matsushita Electric, was reported as making the following comments on the process:

1. All the engineers in all the companies know and talk to each other. There was no 'not invented here' syndrome.
2. MITI's secret was in its committees. A mixture of industry leaders, academics and consumers were selected for its dozens of committees on technological and industrial matters, from restructuring to manned space flight. Through the consensus achieved in this way, areas to be concentrated on were defined.
3. R&D policy is thus a collective one and accepted naturally by all companies.

This, Mr Koratsu said, was the secret of Japanese industrial policy, i.e. MITI performed a coordinating rather than a deciding function.

MITI does, of course, have significant funds for R&D, to spend on what are judged to be programmes with no profit motive, high risk, etc. Energy conservation through solar, nuclear, geothermal and heat pump techniques absorbs by far the majority of the funds. Those subjects are plainly quite different from those which Japan's technological competitors feel need government financial aid. The amount used to finance near market research in industry is not anything like what the Western countries sometimes suppose to be, even then it is a complement rather than a supplement to corporate R&D.

Whatever value the subsidizing of industry in applied research may have, it is not in the relief of the risks and costs of competitive innovations. In general, the contribution of the state is only to the applied research element of the feasibility stage of the total innovative chain. Used in the pre-competitive stage of the development of a new technology, it can be a catalyst in focusing a national industrial sector on the potential for exploitation that it offers. It can also, as we shall see in Chapter 7, act similarly in focusing the national industries of the European Common Market on the opportunities and benefits of a pan-European approach.

The risk – and it is a very real one – is that the political enthusiasm required to sustain the support for the length of time required to achieve results will not last. In Japan – and one detects it to a degree in France – this difficulty is avoided by the politicians leaving the decisions to a consensus of officials and industrialists. Nevertheless, it would seem imprudent of industry to embark on a research programme attracting state subsidy unless that programme fitted into its own product plans and could be continued even if the state subsidy were to be withdrawn.

Direct financing of whole programmes and of companies is another matter altogether and we will consider this in a later section. First, however, we will examine another way in which the technological standing of a nation can be influenced by state intervention.

Offset programmes

The acquisition or securing of a national industrial base through state intervention can take another form in the subsidizing of foreign companies to transfer technology to national enterprises or to set up subsidiaries which should achieve the same result. The bait is a public procurement contract and/or subsidies, tax-holidays, etc., from the regional development funds almost all countries maintain.

At one time the aim was employment, a highly visible political outcome with vote-winning potential. While this is still a factor, increasingly governments are seeking 'technology transfer' in one form or another, believing that transferred technology has a much greater long-term benefit than a head-count of labour employment.

Canada, Australia and India have long-established policies requiring offset in all public sector purchasing. In the case of Australia, a Government Committee was set up in 1984 to examine its effectiveness in what was seen as its prime objective, the acquisition of technologies, and openly reported on how it could be more effective in this respect (Committee of Review on Offsets, 1985). More recently Saudi Arabia and Turkey, amongst many others, have pursued a vigorous offset policy. However, all the major industrialized nations such as the United States, Germany, France, the UK and Japan have demanded and received major offset in consideration of their purchase of equipment for their public services. The UK and France, who have chosen the basic Boeing AWACS system to adapt to their early warning requirements in their defence systems, have demanded and got no less than 130 per cent of contract value promised as 'offset' work for their respective industries. Japan and South Korea, neither short of industrial strength but avid for new technology, have both got substantial technology transfer in exchange for their purchases of defence equipment from the United States and European manufacturers.

Saudi Arabia has systematically used its oil wealth to establish industries capable of competing in the global export market. Sheikh Yamani, in a paper to an international conference in 1983 entitled *The Potential for Industrial Technology in Developing Countries* (Yamani, 1983), explained that the driving force for this policy was the uncertainty of the extractive industries as a long-term basis for the economies of developing countries and asserted that 'the developing countries are more in need of industrialization than the industrialized countries themselves.' The result has been a basic demand of 35 per cent offset in all major Government purchases, with the object of establishing Saudi Arabian-managed competitive enterprises in the long run. The much publicized purchase of the 'Peace Shield' communication, command and control system from Boeing, and the more recent purchase of combat and training aircraft from British Aerospace, has already resulted in significant industrial development in Saudi Arabia.

Under the leadership of its Prime Minister, Tugut Ozal, Turkey has launched a drive to improve its industrial status in parallel with its application for entry into the Common Market. It used its purchase of 160 General Dynamics F-16 fighters – a contract valued at some $4bn – to establish a new aircraft assembly plant with the objective of progressive introduction of local content as production proceeded. Aselsan, the state-owned electronics company, will acquire the ability to make the inertial navigation system. It is intended that other Turkish electronics companies, currently assemblers of foreign-manufactured television and video consumer products, should manufacture components for the F-16 electronics and so move up the ladder of electronic technology.

However, it has also introduced a novel form of venture, particularly applicable to large civil capital projects – the 'build-own-transfer' arrangement. The foreign company or consortium supplying the equipment, in this case electrical power stations, forms a joint venture with a Turkish corporation and builds and operates the plant for a period (say fifteen years) with the agreed objective of recovering the cost of the equipment with interest. At the end of this period the equipment is transferred to a wholly-owned Turkish company.

Subsidized inward investment

While supported from a different source in government, regional policy has a similar interventionist aim – to develop the industrial base of a country – or deprived parts of that country. While it has for many years been a subject for political action inside a country, it has, of recent years, started to play a major part in the spread of foreign multinational companies in the industrialized world, accelerated by the prospect of a true European Common Market in 1992.

The establishment of Japanese subsidiaries in Europe has changed from the time in 1977 when Hitachi hastily abandoned its plans to build a colour television plant in the North of England in the face of outraged protests from local competitors. The merest indication of Japanese interest is now enough to start a stampede of European government and regional officials with highly-refined analyses of the merits of their location and with offers of assistance of every kind. France, the perpetrator of the 'Poitiers affair' in 1982, requiring all Japanese video recorders to be routed through this remote customs post, recently proudly claimed it was attracting Japanese production plants. In the Netherlands, MIP, which is 57 per cent owned by the state, takes equity stakes in US companies on the condition they set up in Holland. This is additional to attractive Government grants towards establishing their facilities there. Austria and several states in the Federal Republic of Germany are similarly in the market with inducements for foreign companies to start up in their territories.

Apart from the UK Government's central 'Invest in Britain' organization, both Wales and Scotland have very active agencies seeking to attract inward investment. The Welsh Development Agency, set up in 1976, has now some 5000 acres of land for industrial development at its disposal and some 20 million sq. ft of factory space occupied by companies attracted to establish themselves in Wales. While this regeneration of an area only recently dependent on trade in coal, steel and shipping has only partly been achieved by subsidized inward investment, there can be little argument that the dynamism and confidence generated by the presence of the foreign companies have contributed to the atmosphere of achievement and growth.

The Scottish Development Agency had the advantage of a significant electronics industrial base in the shape of IBM, Ferranti and NCR dating back to the Second World War. On to this base have now been built some 350 electronics companies in an area which has become known as Silicon Glen, amongst which are such world competitors as Digital, Wang and Unisys. Understandably seeking to do more than attract established enterprises, it has had its setbacks in a partnership in biotechnology. With growth in mind, it financed Damon, a small Massachusetts company with a possible breakthrough in the treatment of tumours of the lymph glands, to expand its activities in anticipation of its success in world markets. However, as is not unusual in such ventures, the product ran into problems in clinical trials and it became clear that the grand expansion plan was totally uneconomic.

Subsidized exports

In 1986, a financial aid package totalling $27m was finalized in Washington to cover the initial costs of a contract won by the US company Control Data Corporation to set up India's first mainframe computer manufacturing plant in collaboration with India's state-owned company ECIL (Electronics Corporation of India). In the words of the newspaper report 'it clinched the order for Control Data against its main rival, Bull of France.' It goes on to say, 'US officials estimate that purchases from the US could amount to as much as $500m', depending on the rate at which local manufacture is built up by ECIL. This was the first sale financed by the US Agency for International Development under the 'mixed credit' programme introduced by the Trade and Development Enhancement Act of Congress.

At about the same time, the US charged Japan with blocking an emerging consensus on reform of tied aid credits to support exports to developing countries. That a consensus should be sought is evidence enough that export credit of this type has become a major form of government subsidy for its exporting industries.

Particularly in the cash-starved Third World, the relative merits in terms of technology, price and delivery of a product or system are less significant than the means of financing its purchase. Buyers then shop around for the best credit terms, Government aid programmes assume great importance and their officials and ambassadors, even their ministers, are drawn into playing a major role in clinching a sale. The name 'crédit mixte' betrays its French origin, although it is now practised widely by all nations and, as the opening paragraphs indicate, the United States has now both recognized the need to use it and the need for some international consensus to keep it within reasonable limits.

The OECD (Organization of Economic Cooperation and Development) has been trying to achieve at least broad rules about its use of an element in aid programmes tied specifically to being spent with the lender country, without success. This has taken the form of a US-backed proposal aimed at raising the minimum percentage of untied (i.e. not specific to purchase from the lending country) from 25 to 40 per cent, so making it more expensive to provide this form of export subsidy. This has foundered on a technicality of how the calculation is to be performed. Whatever the outcome, it still leaves a lot of scope for government intervention and subsidy in support of its exporting companies.

State-financed companies

Britain's pioneering experiment in the 1980s in privatizing state-owned assets in industry and its gathering momentum under governments of all political persuasion and economic status, should not blind us to the continuing existence of a significant body of such enterprises in the world – nor of their importance in the competitive innovation scene. Memories are short. A generally buoyant world economy with a lot of mobile international money is a necessary condition for 'privatization' to succeed. In the years when these conditions did not apply, not a few companies who are major players in the world market today were rescued and only exist today through rescue operations financed in large measure by the state.

Furthermore, where 'privatization' is being pursued, it is a matter of degree, with a complex of ways in which state finance and control are withdrawn from the company. This takes the form of the percentage shareholding with certain European companies claiming to be in the private sector when a substantial proportion, even a majority, of the shares is held by the state. Even in probably its purest form, in Britain in the 1980s, the state has retained what is called a 'golden share' and through this retained control over any change of ownership. Since this, in the view of many, is the essential and ultimate test of a private sector company – that it can be taken over – this is no trivial residue of state control.

However, whatever the degree of state finance involved, there can be no question that this form of intervention by the state is still alive and well and must be taken into account for the future. Companies such as Volkswagen, Renault, CGE, Thomson and Rhône-Poulenc listed in Table 3.1, all have a substantial element of their equity provided by the state or state-controlled institutions. Neither their behaviour nor their success in global markets conforms to the image of dull, unenterprising bureaucracies which state enterprise is supposed to be. To examine this further let us look at the French approach to state participation in industry.

The undoubted success that France has achieved in the use of state intervention in industry generally is attributed to its special class of people, the Noblesse d'État, carefully created and nourished through their careers by the system of the 'grandes écoles'. While many countries, and Britain in particular, have their systems for recruiting and developing professional government administrators, only France has developed this into a prime source of top management of both state and private sector enterprises. The selection process for this Noblesse d'État by-passes its universities, going from the top academic lycées, such as Henri IV, through to the Napoleonic École Polytechnique, or the École Normale Supérieure, etc., and thence to the École Normale d'Administration. The end product then enters one of the grands corps such as the Inspection des Finances, the Cours des Comptes or the Conseil d'État and so up the ladder of government services.

However, it does not end there. It has been estimated that some 75 per cent of the chairmen of France's top 100 companies came from this source, as well as all the big names in French politics – Chirac, Giscard d'Estaing, Fabius, Rocard, etc. When it comes to the consensus of purpose necessary for state intervention to work, clearly France has a formidable lead.

What would be unthinkable for nationalized companies in other countries is commonplace in France. Pechiney – an aluminium smelting enterprise – got into trouble in the early 1980s and was nationalized. First under Besse (who was moved to Renault) and subsequently under Pache it has invested heavily in Canada and Australia, has diversified into high added value speciality materials and has a wholly-owned Japanese subsidiary with approaching £200m sales. Pache is a graduate of the École Polytechnique and École des Mines.

Rhône-Poulenc, again a state-owned French company, hit the headlines with its massive global acquisition activities. Again, the objective has been to get out of bulk chemicals into the high value added (synonymous here with innovative) sectors such as agrochemicals, pharmaceutical and fine chemicals. On the long list of some fifteen major acquisitions outside France are, for example, the agrochemical business of Union Carbide in the US and the fine chemicals business of RTZ in the UK.

Even Renault, only just emerging from a dismal profit record and massive state subsidies, has been active in acquisitions in the US market and is in the process of creating an alliance with Volvo (as discussed in Chapter 3).

All this means raising lots of money, taking risks and having long-term strategic plans. This can be contrasted with the experience of nationalization in the UK. A report commissioned by the government in 1975, around the time when the Bill nationalizing the aircraft and shipbuilding industries was going through the British Parliament, opens with these statements (NEDO, 1976):

> There are certain features of the relationship between government and nationalized industries which came through so clearly in our enquiry that we believe they can be stated without risk of contradiction:
> - There is a lack of trust and mutual understanding between those who run the nationalized industries and those in government (politicians and civil servants) who are concerned with their affairs.
> - There is confusion about the respective roles of the boards of nationalized industries, Ministers and Parliament, with the result that accountability is blurred.
> - There is no systematic framework for reaching agreement on long-term objectives and strategy, and no assurance of continuity when decisions are reached.
> - There is no effective system for measuring performance of nationalized industries and assessing managerial competence.

It is not difficult to see the manifestation of the great cultural gap between the UK and France in their approach to nationalized industries. The report itself goes on to acknowledge this difference, which it attributes to a 'consensus' rather than 'arm's length' approach of the State to its industrial enterprises. Others may call it 'dirigisme' but the report itself sees this as a by-product of France's unique élite interchangeable between industry and government. We shall return later to this consensus which, by different routes, is also achieved in Germany and Japan.

The United States, like Britain, does not seem to be able to achieve this form of consensus. A group of US business leaders recently launched a 'Wake Up America' campaign, aimed at creating an industry/Government partnership to revitalize US manufacturing and restore the country's lead in technologies ranging from machine tools to high definition television. James Koontz, chief executive of Kingsbury Machine Tools in New Hampshire and Chairman of the National Center for Manufacturing Sciences consortium, is quoted as saying his industry needed $100m to develop the next generation of equipment (*Financial Times*, 1989), when the entire market capitalization of the industry was only $600m!

These industrial cultural differences come out in our final examples of government intervention in innovation – the unique Airbus Industrie enterprise and the present and future development of space.

Airbus Industrie

In the 1950s, the rapidly developing air transport industry was served by many aircraft manufacturers – Vickers, de Havilland, Bristol, Sud Aviation, Lockheed, Douglas and Convair amongst others. All these were subsidized by the state but, suddenly, in a world converted to the merit of jet transport by the de Havilland Comet, appeared the long range (by the standards of the day) Boeing 707. Undoubtedly given tremendous launch aid through the US Air Force development and production order for its predecessor the KC-135, Boeing brilliantly exploited this and captured the bulk of the long-range market (but see Chapter 2 for Boeing's subsequent difficulties).

France (industry and state) had tasted success with the medium range Caravelle but, conscious that it could not match the massive KC-135-based launch of the Boeing company, and at that time absorbed in thoughts of 'le défi américain', set about creating a competitive airliner – the Airbus – on a pan-European basis. To bring the partners together, it proposed the Groupement d'Intérêt Économique (GIE), established under French law, to give some status to French companies coming together to collaborate in new technology.

Today the partners in Airbus Industrie are Aérospatiale (37.9 per cent), Daimler-Benz (MBB until recently, 37.9 per cent), British Aerospace (20 per cent) and CASA of Spain (4.2 per cent). Of these Aérospatiale and CASA are nationalized and, up to its acquisition by Daimler-Benz, MBB had a large part of its equity in public or quasi-public hands. It is not irrelevant to note that Daimler-Benz demanded, as a condition of acquiring MBB's responsibilities in Airbus Industrie, that the German Government shield it from losses through movements in the DM/US dollar rates of exchange.

Some twenty years on, Airbus Industrie has established itself as a producer of innovative and competitive airliners – so much so that the United States, no doubt prodded by Boeing, is pursuing a vigorous attack on the unfair competition it is able to offer through the benefits of substantial state subsidies. Unease about this continuing indefinitely has been expressed by the UK Government and in the body advising the Bonn Government on economic affairs. A committee of 'wise men' was set up to chart a way of this GIE turning itself into an international company capable of raising its future financial requirements in the capital markets. A finance director has come – and gone – and there is talk of replacing the work division on the aircraft by preordained percentages to a competitive style of subcontracting.

In respect of our study, it must be recognized more as a survival of an outdated mode of international collaboration than pointing the way to the future. In the climate of the 1960s it had to preserve the concept of national champions and award work to them in proportion to their state's contribution to the costs. Many of its current difficulties are attributable to this. Nevertheless, however easy it is to criticize the arrangement with hindsight, it can justly claim to have given the world better airliners by challenging, with new ideas, Boeing's near monopoly of this market. How, one must ask, could it have been done in any other way, particularly if it were left to the private sector to mount the challenge?

Space programmes

This is a classic case of the birth of a new, potentially important, macro-innovative sector. Like the situation when the Wright brothers took their first tentative venture in manned flight, writing the commercial prospectus for investment verges on the impossible. Unlike the Wright brothers situation – and akin to the harnessing of nuclear power – it requires enormous sums of money. The establishment of a space station, the estimates for which exceed $20bn, has caused even the United States Government to seek the participation of European governments as well as of Japan and Canada.

The accepted wisdom in all countries has been that the overall direction of man's new-found ability to use the space outside this planet must be managed and financed by governments. The leaders in the early days, the United States and the USSR, were clearly of this mind and the European Space Agency (ESA) and the newcomers in Japan and China, capitalist and socialist economics apart, are of a like mind.

This is now beginning to change. Arianespace, the French company which sells satellite launches on the French Ariane launcher, recently announced that it planned to place an order for fifty Ariane-4 (the latest version) launchers for delivery between 1991 and 1997. Arianespace has a majority of private shareholders with only 34 per cent held by the French Government. There is talk – no more – in the United States of the private sector putting up the money for launches using the existing NASA facilities for the actual launching process.

In the meantime the process of exploration for exploration's sake goes on in the tradition of the Apollo programme to put man on the moon. Nevertheless, this time the USA is seeking international partners for a permanent space station in orbit round the earth. In the UK, fired with concern that there was no foreseeable end to the frivolities that could be proposed in the name of space exploration, the Government in the 1980s sought to abandon it to the private sector. Such an investment, and the risks involved, could only be faced in the private sector by a global private sector enterprise exceeding by orders of magnitude the scale of the largest of today's macroindustrial companies, drawing on world level financial markets with resources matching the gross national product of a major nation. With even Eurotunnel in financing difficulties, such a prospect seems remote in the extreme!

Industrial policy

The shortcomings of certain forms of state intervention in industrial affairs – and in its most important component innovation – have been used to make the case for the state having no need for an industrial policy. This is nonsense.

It is nonsense, first, for the fundamental reason that politicians have no justification for their existence if they admit to being impotent in the face of the 'hidden hand' of market forces. They are elected to 'do something' for the wellbeing of the nation and, as is universally acknowledged, wellbeing is meaningless without the wealth creation needed to finance it.

It is nonsense because, in international trade, the state must represent the interests of its industrial sector in the councils of nations. Even dedication to the promotion of free trade and the minimization of state intervention is an objective which is as meaningless as 'disarmament' unless it is accompanied by detailed tactical policies to develop and police it. Indeed 'free trade' requires a much more detailed industrial policy than the simple doctrines of protectionism.

It is nonsense because there are certain things that only the state can do – education and basic research at one extreme and macroinnovative ventures at the other – for which it must have an industrial policy framework. The accords on which world trade is regulated must be underpinned by an international consensus on industrial policy derived from the proper representation of national interests – another way of describing a national policy.

What seems to emerge from the success stories of national policies is that they must be the result of a consensus of industrialists, officials and representatives of the social aspects of national life which transcends the political in-fighting of democratic government. Wealth distribution may well be the proper province of politics, but wealth creation, particularly in so far as it depends on innovation, cannot be pursued as a short-term political game. This has always been true in national economic life and it is an inescapable imperative for life in the global marketplace. History may well prove that vote-catching, politically-oriented national policies for industry have done more harm than good. Nevertheless, a country without a consensus on its industrial policy enters the competitive global marketplace of today at a most serious disadvantage.

References

Alvey Directorate (1986a), *A Plan for Concerted Action*, IT Committee report, DTI, London.
Alvey Directorate (1986b), *Alvey Programme Annual Report*, DTI, London.
Coleman, R. (1989), *Industrial R&D and Public Policy – a UK View*, DTI, London, paper dated 31 May.
Committee of Review on Offsets (1985), Australian Government Publishing Services, Canberra.
Financial Times (1989), 'US Business Seeks Public Aid.' *Financial Times* Report, 26 May.
Finniston Committee (1979), *Engineering our Future*, HMSO, London.
Grindley, P. (1987), London Business School Release, *Financial Times*, 23 November.
Henderson, D. (1986), *Innocence and Design*, BBC Reith lectures.
House of Commons (1988), Report of the Committee of Trade and Industry, HMSO, London.
Japan Technology Transfer Association (1987), March, Tokyo seminar.
Jones, A. (1985), *Britain's Economy – the Roots of Stagnation*, Cambridge University Press.
League of Nations (1933), 'The State and Economic Life', International Conference, London.
MOD (1983), *Value for Money in Defence Equipment Procurement*, UK MOD Defence Open Government Document 83/01.
National Audit Office Report (1988), May, HMSO, London.
National Economic and Development Office (1976), *A Study of UK Nationalised Industries*, HMSO, London.
Okimoto, D. T. (1989), *Between MITI and the Market*, Stanford University Press.
Yamani, H. E. (1983), *The Potential for Industrial Technology in Developing Countries*, Man and Technology Symposium in London, published by Cambridge Information and Research Services.

7 The European Community policy for innovation

- The European challenge
- Information technology and telecommunications
- The EC initiatives
- Expenditure on R&TD
- The 'framework programme'
- Intention and opportunity
- Scope for intervention
- Conclusion

As argued earlier in this book, innovation on a macro scale represents a serious challenge to all those who aspire to become major players on the industrial scene. The impact of rapid technological developments on the wellbeing of modern enterprise was demonstrated by many examples in the preceding chapters. It is important in this context to review the effect of policy in the European Community (EC), which is already established as one of the major economic power blocs, since this policy is likely to affect the future of most enterprises in the Community and elsewhere.

The European challenge

The creation of a single market in Western Europe, starting with the abolition of trade barriers in 1992, signals an important potential shift in the balance of economic power in world trade. As many realize, the Single European Act, which aims to open the internal market in 1992 to all member states, is only the beginning of a fundamental realignment process with many political and social ramifications. At the root of it is the restructuring and reorganization of European industry, not only to meet the internal needs of the market, but to become a significant economic force on the world stage, against the stiff competition of the two other power blocs – the USA and Japan. At a time when industry is overwhelmingly affected by rapid changes in technology, the EC is faced with the need for determining a coordinated policy for science and technology (S&T) or research and development (R&D), in order to equip industry with the tools needed to thrive in the future.

This challenge has been fully recognized by the Commission of the European Communities (CEC) as is evident from its series of reports on the subject and from a substantial S&T budget, which is intended to steer the activities of member states and industrial companies in this area. The question is whether proclamations of intent are sufficient to achieve the goals of the CEC and whether the budget would become an instrument of central direction and control. A further question is whether the CEC policy and the creation of a large central budget will help or threaten the innovation process of private enterprise. The purpose of this chapter is to review the current state of the proclaimed policy and to highlight some of the problems and the consequences that may ensue. Throughout this chapter the reader should bear in mind the delineation of the subject matter, and in particular the relationship between science and technology on the one hand and research and development on the other, which are discussed at some length in Chapter 1.

Information technology and telecommunications

The area of telecommunications highlights one of the main directions in which the EC policy on R&D is evolving. In a document entitled *Telecommunications: The New Highways for the Single European Market* (CEC, 1988), the Commission of the European Communities – as pointed out in Chapter 3 – remarks on the fast growing world market for telecommunications services, estimated to be worth well in excess of 500 billion ECU (1 ECU equals about 1.1 $US), and recognizes the two options for Europe – either to operate on equal terms at the global level of the world economy, or to become a second-rank partner and grow steadily poorer. Interesting extracts from the CEC report, highlighting the size of the market and the scale of investment required for innovation in this field, are given in Chapter 3, and the document further states:

Nor can the Community be content with a strategy that is only industrial and regulatory. To bear fruit, the joint European effort must have a favourable environment. This means that it is equally vital:

- to manage the social effects of technological change, which will create many jobs but will do away with others;
- to determine a common approach to access to information, protection of private life, etc., so that the individual feels at ease in the new technological environment;
- to take account of the interests of less-favoured regions, so that moderniz-

ation of telecommunications helps to increase the cohesion of the Community rather than widen the existing gaps.

The challenge for the future is clear, and this explicit summary of the CEC's objectives contains a framework for a series of initiatives planned by the Commission, briefly described below.

The EC initiatives

To take up this challenge, the EC has embarked on several initiatives:

1 Various agreements and action programmes were signed during the 1980s to remove all internal trade barriers and thereby open up the telecommunications market in order to set the scene for closer collaboration between the major operators in the sector.
2 It is widely recognized that 'the national system of standards in force differ considerably from each other, thus depriving industry in the Community countries of the benefits of economies of scale' (EC, 1988). An ambitious and wide-ranging programme has, therefore, been undertaken to introduce technical standards in the industry, with the aim that all future development of hardware and software would conform to the standards and achieve compatibility of equipment, design and operating procedures that are the envy of any highly fragmented industry. After the abolition of trade barriers in Europe, *standardization is perhaps the most significant and far-reaching action taken by the EC to open up the market*. It will bring about a revolution in the way companies formulate their product strategies and plan their assault on world markets.
3 As an example of the effect of standardization, it is estimated that the creation of common market technical standards for terminals, together with uniform testing procedures in approved laboratories (to replace the twelve national procedures in existence), would save the EC about 1 billion ECU per annum.
4 Another outcome of standardization would be 'compatibility and interoperability of systems; common tariff principles, to ensure reasonable, market-led prices; development of European high value-added services.' Armed with all these advantages, Europe should be able to fend off foreign competition and win sizable orders abroad.
5 In line with these ambitions, the EC is intent on formulating an extensive programme for developing joint research. As the CEC document, mentioned earlier, proclaims (CEC, 1988):

> Upstream from marketing and standardization, a major research effort is required to develop the technology of the year 2000. The Community's

RACE programme is intended to enable the necessary technology and standards to be perfected for the future broadband integrated network: high-speed and high-complexity integrated circuits, integrated optoelectronics, broadband switching, passive optical components, high bit-rate links, concepts for system development and integration, etc. The RACE programme is based on cooperation between industrial laboratories and universities all over Europe and has a budget of 1200 million ECU, financed by equal contributions from the Community and industry. Other Community programmes also encompass certain aspects of telecommunications: Esprit II (European Programme for Research and development in Information Technology), AIM (Advanced Informatics in Medicine), Delta (Developing European Learning through Technological Advance) and Drive (Dedicated Road Infrastructure for Vehicle safety in Europe).

6 Other initiatives of the CEC will involve closer harmonization of company law, taxation, terms of employment, and alignment of policies relating to competition throughout the EC, including fair trading and the treatment of monopolies and mergers within and across national boundaries.

Expenditure on R&TD

In the EC document entitled *Research and Technological Development Policy* (EC, 1988), the increasing importance of science and technology in the world is strongly indicated. The dynamic and expanding nature of this activity is highlighted by the fact that some three million scientific papers are published every year in specialized journals and it is estimated that some 90 per cent of all the researchers and scientists throughout human history are alive today!

A large proportion of this human scientific and technological resource resides in the three main economic blocs: USA, Japan and the EC. It is estimated that there are about 720,000 research workers in the USA and some 440,000 in Japan; whether the EC with a population of 325 million is adequately equipped for the challenge with some 450,000 research workers is open to question, though comparisons in this field are notoriously difficult, since definitions of scientific and technical qualifications vary greatly from country to country. Be that as it may, the EC is well behind in the race to dominate the major industrial sectors of the world economy: 'of the thirty-seven technological sectors of the future that have been identified, thirty-one are dominated by the USA, nine by Japan and only two by Europe ... (some sectors being dominated by two countries equally)' (EC, 1988).

These figures are obviously a cause for concern, as the EC document makes clear:

> Why then does Europe make such poor use of the immense scientific, human and economic potential that it appears to possess? There are many reasons ... insufficient attention to the problems involved in the follow-through from scientific research to technological development and then on to the marketplace; poor marketing techniques: the persistence in companies of organizational principles and training methods unsuited to the circumstances of the scientific and technical revolution; the lack of financial instruments to stimulate R&TD and the meagre venture capital available for investment in this field. All these factors will continue to constitute a drag on European R&TD until there has been a radical revolution in attitudes, practices, behaviour, culture and education in the Community (EC, 1988).

The EC prefers the term R&TD (Research and Technological Development) to R&D, in order to emphasize the important contribution of technology. 'Economists realized long since that R&TD play a vital role in the process of economic development. Used wisely, science and technology can also help enormously to increase general wellbeing and improve the quality of life for individuals and society: science itself enables us to understand and correct the accidental harm done to the environment and human health by certain technical activities' (EC, 1988).

Total expenditure on R&TD (public and private, civil and military) is enormous and growing. As shown in Figure 7.1, the expenditure in 1985 for the twelve countries of the EC amounted to 65bn ECU, which though admittedly higher than that of Japan, was less than half of that spent in the USA. It is perhaps not widely acknowledged that the level of scientific activity in a country does not have an immediate relationship either to its expenditure on R&TD or to its economic success. If scientific excellence is measured by the number of Nobel prizes awarded, it is notable that from 1950 to 1987 the number of prizes gained by the USA, Japan and Europe were 115, 4 and 86 respectively. It has, therefore, been suggested that producing Nobel prize laureates is indeed a cause for immense national pride, but that they make only a minor contribution to increasing the gross national product. Expenditure on R&TD, however, is thought to be a better indicator of the level of resources invested to improve economic wellbeing. In that respect, expenditure on R&TD in Europe is generally thought to be far from adequate.

What is even more worrying to many commentators is that the rate of growth of this expenditure in Europe is relatively low. Between 1981 and 1985 the expenditure rose from 45bn to 65bn ECU, an increase of 44 per cent, whereas it more than doubled in Japan and the USA during the same period. Furthermore, 'research intensity' (R&D expenditure as proportion of GDP) is only 1.9 per cent in the EC, compared with 2.8 per cent in the USA and 2.6 per cent in Japan (CEC, 1989a). The report strongly

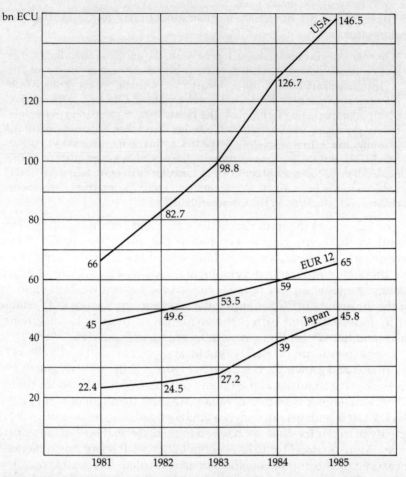

Figure 7.1 *Estimate of gross domestic expenditure on R&TD (public, private and military expenditure) from 1981 to 1985*
Source: OECD

argues that 'Europe as a whole is spending less on R&D than its major competitors; and European industry much less. Moreover, the share of research scientists and engineers in the labour force is significantly higher in the USA and Japan.'

Another important issue is that of defence R&D: 'The high level of defence R&D in the USA, half of which is contracted out to industry, is a potent factor in mobilizing resources. The precise level of spin-off to the civil sector from military R&D in general, and from SDI in particular, is hard to assess. But there have been undoubted major benefits in the fields of materials, aircraft engineering, vehicle technology, computer technology and systems engineering. The shift during the 1980s of much military R&D towards development work tailored to the needs of more fundamental and applied research has, however, raised some concern in the USA about reduced industrial impact' (CEC, 1989a).

A further concern in the USA is 'the prospect of shortfall of perhaps 500,000 scientists and engineers by 2010 because of demographic trends and the pattern of university enrolment.' As a result, there are already signs of American firms and institutions trying to entice foreign students to remain in the USA after graduation, and special encouragement is given to post-doctoral students. Developing countries are particularly affected by this trend, since they are likely to lose their most talented scientists and engineers to the USA, but recruitment of European personnel has also intensified in recent years and this continuing brain drain will become increasingly serious when the main protagonists continue to vie with each other for skilled manpower in the fields of technological innovation and management.

Not surprisingly, the expenditure varies greatly from country to country within the EC, as shown in Figure 7.2, and the disparities between the various member countries cannot be explained just by the respective size of their populations. The main effort comes from three countries (Ger-

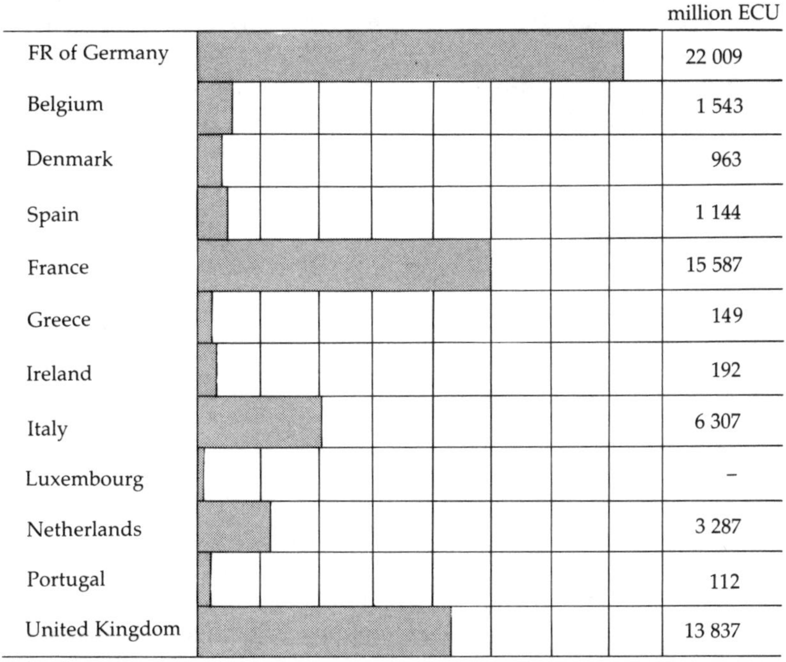

	million ECU
FR of Germany	22 009
Belgium	1 543
Denmark	963
Spain	1 144
France	15 587
Greece	149
Ireland	192
Italy	6 307
Luxembourg	–
Netherlands	3 287
Portugal	112
United Kingdom	13 837

Figure 7.2 *Estimate of gross domestic expenditure on R&TD (public and private, civil and miltary expenditure) in 1985*
Source: OECD

many, France and the UK), who between them account for some three-quarters of the total spending on R&TD, but only Germany has achieved a 'research intensity' (2.8 per cent) comparable to the EC's main competitors (CEC, 1989a). The range in 'research intensity' among member states is very wide, being of the order of 12:1 at a national level, while at a regional level the disparities are much wider and reflect great variations of economic wellbeing across the EC.

Altogether, the effort in Europe has hitherto been 'unbalanced and fragmented: funds spread too thinly, research teams working in ivory towers, a lack of coordination, poor dissemination of information, inadequate mobility of research scientists, duplication in national programmes, differing strategies, disparities in standards' (EC, 1988), in short: the lack of a single open market. Although a great deal of effort has been made to improve the situation through a large number of cooperative transnational programmes, such as BRITE (Basic Research in Industrial Technologies for Europe), CERN (European Centre for Nuclear Research), ESPRIT (European Strategic Programme for Research in Information Technologies); EURAM (European Research on Advanced Materials); ESF (European Science Foundation), ESA (European Space Agency), ESO (European Organization for Astronomical Research in the Southern Hemisphere), and many others (these programmes have been discussed by many writers, see for example Audretsch, 1989), they still account for a small percentage of the total research effort (CEC, 1989a). The purpose of the Single European Act is to create the conditions which will allow all these deficiencies to be eliminated.

The 'framework programme'

In response to the need to formulate an effective policy for R&TD in the EC, the so-called 'framework programme' for 1987–91 was published, listing the eight main areas of activity deemed necessary for the Community to undertake, as shown in Table 7.1 (CEC, 1989b; EC, 1988).

As these figures suggest, the major designated activity (over 42 per cent of the total) is that of information technology and telecommunications, which is the fastest growing industrial sector in the world. Bearing in mind that the Community's trade deficit in electronics alone was about 15bn ECU in 1986 and that European manufacturers hold about one-quarter of the world market in information technology and telecommunications, the sector clearly poses an immense challenge, with undoubted potential rich rewards to the successful, but coupled with enormous risks. It has been estimated that the cost of developing a new digital switch can amount to $700–1,000m. As the amount of resources devoted to research in the telecommunication equipment industry is normally

Table 7.1 *The EC framework programme (1987–91)*

		m ECU	%
1	Quality of life (health, radiation protection, environment)	375	6.9
2	Information technology and telecommunications (including transport)	2,275	42.3
3	Modernization of industrial sectors (manufacturing, advanced materials, raw materials and recycling, technical standards)	845	15.6
4	Biological resources (biotechnology, agroindustrial technologies, agriculture and competition)	280	5.2
5	Energy (fission, thermonuclear fusion, non-nuclear)	1,173	21.7
6	Science and technology for development	80	1.5
7	Marine resources (marine science, fisheries)	80	1.5
8	European science and technology cooperation (human resources, major installations, dissemination of results)	288	5.3
	Total	5,396	100

estimated to be about 7 per cent of sales, it follows that the level of sales needed to cover the development costs is about $14bn, which in the mid-1980s was bigger than the size of the largest European market. These figures indicate the degree of challenge and risk involved and underline the discussion on this issue in Chapters 2 and 3.

Arguably, the initiative for a concentrated research effort in a bid to become a global market player should have been launched much earlier, since there is always a danger that aspirant late-comers may not be able to catch up with leaders. But a bold initiative on such a grand scale 'on behalf of Europe' could not be imposed without the acquiescence of member states and therefore had to wait for a favourable political climate. Also, technology in this area is moving so rapidly that the rate of obsolescence (and the resultant cost) can be crippling if huge investments in R&TD are not recovered by sales volumes, so that the strategy of being first has its own risks, and industrial history is full of examples of defeated pioneers, who achieved no more than paving the way for their successors. Balancing these risks is the essence for an initiative of this kind and timing can be crucial to its ultimate success.

The second largest item in the 'framework programme' (amounting to about 22 per cent) is that of energy, reflecting the general concern that Europe is highly dependent on oil from Third World countries. Low cost energy is not only desirable for domestic consumption, but is essential for industry in order to keep down production costs. Amelioration of the dependence on imported fuel – by research into new sources of energy (such as controlled nuclear fusion), by exploring the use of renewables, and by a far-reaching programme of conservation and better use of energy – is a high priority objective for the EC.

Almost 16 per cent of the budget is allocated to modernization of industrial sectors. Bearing in mind that manufacturing industry in the EC accounts for 30 per cent of the GNP and employs some 41 million workers (about 75 per cent of the industrial workforce), there is clearly a great deal to be gained from making industry more cost and value-added conscious. Introducing widespread standards, exploring the use of new materials and new manufacturing processes, improving quality control, and developing advanced control methods for integrated and automated manufacturing systems – all these initiatives are designed to contribute towards that goal.

Although each of the remaining items in the 'framework programme' constitutes a much smaller percentage of the budget than the top three, the sums of money involved are quite substantial. They all aim to achieve improved collaboration and coordination, a more purposeful direction of effort, more efficient dissemination of results, and better diffusion of technology. In short, the prime objective is to make Europe more competitive on the world scene, both in relation to the two main economic power blocs, the USA and Japan, and to the newly-emerging industrial countries in the Pacific Basin.

The CEC is conscious of the fact that the 1987–91 'framework programme' is only the first ambitious step forward and it has already announced (CEC, 1989b) a provisional follow-up programme for 1990–4, shown below. This programme, which is to be finalized in due course following discussions by the Council of Ministers and the European Parliament, continues the theme set out in the first programme, as shown in Table 7.2.

As can be seen from the second programme, which has a proposed budget of about 43 per cent above that of the first programme, the items under the heading 'enabling technologies' continue to dominate the EC policy for R&TD and account for well over half the total budget. The absolute amount allocated to energy is marginally below the corresponding figure for the first programme, while the budgets for life sciences and the environment are substantially increased. A notable addition to the second programme is that of 'management of intellectual resources', reflecting the Commission's desire to see freedom of movement of scientists and engineers across national frontiers within the EC.

Table 7.2 *The proposed framework programme (1990–4)*

		m ECU	%
I	*Enabling technologies*		
	1 Information and communications	3,000	38.9
	2 Industrial and materials technologies	1,200	15.6
II	*Management of natural resources*		
	3 Environment	700	9.1
	4 Life sciences and technologies	1,000	13.0
	5 Energy	1,100	14.3
III	*Management of intellectual resources*		
	6 Human capital and mobility	700	9.1
	Total	7,700	100

Intention and opportunity

The purpose of the two 'framework programmes' and other CEC documents is to impress upon the captains of European industry the urgent need for innovation through investment in R&TD. This exhortation is put forth both by exposing European industries to unfavourable comparisons with their American and Japanese counterparts and by offering a central budget as a financial foundation on which further industrial expenditure on R&TD can be built.

What is somewhat surprising is that the CEC documents do not make a clear distinction between micro and macroinnovation. From our earlier discussions it would appear that such a distinction is important, in that while microinnovation may be regarded as the province of individual companies in relation to their existing product range, the purpose of an overall R&TD policy is to create a framework for grand alliances at the macro level. Admittedly, many of the research topics enumerated by the CEC are of a long-term nature and would involve massive financial resources, which individual companies would find difficult or impossible to provide on their own. Equally, however, research projects are listed that could impinge on microinnovation programmes that individual companies could wish to undertake themselves.

It may, therefore, be argued that the distinction between micro and macroinnovation in the 'framework programmes' is blurred, and in that sense the CEC policy and its budget are (or can become) more interventionist than originally intended. Also, the EC budget projections are more in the nature of a shopping list of desiderata, and the broad descriptions of some of the research topics provide only an overall sense of direction, whereas a macroinnovation strategy would have to be more selective, more structured, and more specific about its objectives and industrial participants. Perhaps such a programme can only emerge through the initiative of the would-be industrial participants themselves.

Becoming more competitive is not the only objective of the EC Commission, as is clear from the following statement:

> Science and technology are vital in improving Europe's competitive position in the context of the completion of the internal market. They also have a major contribution to offer in meeting the societal needs of the European society for a cleaner and safer environment and better health care (CEC, 1989a).

A better environment, improving health and enhancing safety are explicitly mentioned as areas that S&T (or R&TD) can contribute to. But in a sense these are natural expectations that, in the current climate, are bound to follow from an overall improvement of the economy in Europe. They are laudable aspirations, calculated to appeal to many interested parties, including politicians, industrialists and trade unionists. While many may be somewhat uneasy about the size of the CEC budget for R&TD, fearing the dead hand of central bureaucracy and supra-state intervention, it is clear that with insufficient financial resources the CEC would be unable to achieve swift reforms on the lines indicated earlier in this chapter. The CEC thesis is that the central budget would become a magnet for industrial collaboration, so that a consensus would emerge regarding the desirability for standardization, for elimination of fragmentation and duplicated effort, for European patents and for quality testing procedures to replace national conventions and traditions.

Scope for intervention

As intimated earlier, it should be appreciated that the 'framework programmes' provide the Commission with a hidden agenda for political and social change. For a start, the open market is expected to increase competition, which in turn would encourage firms to increase their expenditure on R&TD, in order to retain or improve their relative position in the market. The general expectation is that increased competition would make firms more efficient, to the benefit of consumers in terms of price and quality. But concern has been expressed by some industrialists that increased R&TD expenditure would instead increase total costs, so that the cost of the final product to the consumer may consequently rise. This view is rejected by most pundits, who expect the savings achieved by increased productivity and industrial efficiency to more than compensate for the increased expenditure on R&TD.

As indicated in earlier chapters, it is probably inevitable that small companies may be at a disadvantage in the free market and would not be able to afford either to increase their R&TD spend or to invest in modernization of their operations. This means that the trend for joint ventures, and particularly mergers and takeovers to create larger industrial and commercial firms, would continue and even accelerate for a while. For small and medium-size firms currently spending little on R&TD this trend could well spell eventual extinction, or at least loss of independence. It is, therefore, argued that the CEC programmes are bound to favour current and future big firms. Effective measures, therefore, will have to be taken to avoid the emergence of monopolies, since monopoly power discourages innovation and kills competition. Clearly, that would be detrimental to consumers and consequently to the long-term interest of European industry in having to face the harsh realities of global competition.

There are other possible scenarios that have been postulated from the development of a free and open market in the EC: free mobility of manpower could lead to skilled personnel gravitating towards areas 'where the action is' and this drain could have serious consequences for poor and remote regions; against that, production of high value-added products would tend to move to low labour cost regions, in a way reminiscent of the migration of plants in the USA from the north to the south. Whether the migration of manufacturing capacity would cause a similar migration of industrial R&TD to poorer regions in Europe is debatable.

It is clear, though, that the CEC can affect the siting of multinational research centres created or heavily supported by EC funds and that policy makers would be tempted to make such siting decisions on social grounds, in order to reduce the wide economic disparities between various regions within the Community. Decisions of this nature may not be compatible with the long-term need to site facilities and generally allocate resources for the main purpose of making European industry competitive in relation to the USA and Japan. Although this fundamental conflict between the economic and social objectives of the CEC must have been fully appreciated by the policy makers, there is no indication that a rational machinery has been set up to resolve it. Like many other issues in the EC, there is always the danger that macro decisions involving long-term consequences would be settled on a tit-for-tat basis, following prolonged political wrangles fuelled by national interests.

The role of public procurement adds another dimension of central intervention and control. Leaving aside defence procurement and defence R&TD, which are bound to exercise a profound influence on European industry (in the same way that they have, for many years, in the USA), public procurement can take the form of overt or covert state aid to specific industries or geographical regions. It can even become an instrument for 'controlled privatization', where tenders are issued to

undertake particular public services on closely defined terms and conditions, for example to rid a town of pollution, or to plan and maintain certain transportation services. Such projects have a strong R&TD element, but their terms of reference could be extended to include implementation and management.

Conclusion

Thus, the evolution of the current policy has the scope of producing a curious mixture of an open and free market, coupled with a potential for central control and intervention on a massive scale. Both industrialists and policy makers need to be aware of the contradictions inherent in this scenario and the possible dangers of impeding the process of free enterprise, particularly in the area of macroinnovation. The vast financial and manpower resources needed for 'big science' and for macroinnovation means that a central policy with the attendant budgets would continue to have a vital role to play, but as the future of the EC unfolds, the powers-that-be have to be fully prepared to see a diminution in the power of the centre to ensure that the free market is offered the necessary conditions to thrive.

References

Audretsch, D. B. (1989), *The Market and the State*, New York University Press.
Commission of the European Communities (1988), *Telecommunications: The New Highways for the Single European Market*, October.
Commission of the European Communities (1989a), *The First Report on the State of Science and Technology in Europe*.
Commission of the European Communities (1989b), European Community Research Programmes.
European Communities (1988), *Research and Technological Development Policy*, Office for Official Publications of the European Communities, Luxembourg.

8 The multinational scene

- The globalization of industry
- The challenge to sovereignty
- Competition policy, mergers and acquisitions
- Taxation
- Regulation of trade – the General Agreement on Tariffs and Trade
- The globalization of sovereignty

The globalization of industry

In the preceding chapters we have set out the compelling reasons why industry, at all levels, must seek the 'critical mass' required to compete on a global scale. At the industrial and microindustrial level this may be achievable from a national base with no more than a sales presence in the key foreign markets. At the macroindustrial level, where the need for market leadership is the driving force, it may well mean establishing a full complement of industrial activities from new product design to after-sales service and support.

The realities of this macroindustrial scene and the problems it poses for industry are comprehensively examined in a collection of studies of the Harvard Business School entitled *Competition in Global Industries* (Porter, 1986). The Introduction and Summary concludes with the unequivocal statement:

> Global competition is no longer a trend but a reality. Today, global competition can rarely be dealt with simply through exports or with free-standing foreign subsidiaries. It must be met through coordinated global networks of activities. Understanding global competition will be the difference between success and failure for many firms and some governments.

The 1980s has been a decade of intensive globalization, features of which have already been presented in earlier chapters. The fact that it is not an easy and certain course of action has not deterred companies from following this course. Takeovers can be disappointing, joint ventures collapse and the cost of establishing 'stand-alone' new subsidiaries in foreign countries painful. Nevertheless, the strategic importance of globalization is such that these risks have to be faced.

Japanese companies are prominent in pursuing this strategy, with companies such as Hitachi and Fujitsu, Nissan and Toyota making the headlines in their pursuit of a European presence. French state-financed companies have become very active in recent years – so much so that the problem of the propriety of the French state having a controlling interest in companies outside France is becoming a serious issue. The process of globalization is further encouraged by those advisers to governments concerned with macroinnovation. By encouraging multinational private sector financing of such projects, the creation of appropriate transnational enterprises to carry them out is an inevitable consequence. Eurotunnel, the project to build and operate a tunnel linking the UK and France, was deliberately financed on a global basis and has turned to banks from all over the world for the money required to meet the majority of the launch costs.

In the new situations created by the collapse of the command economies of the Communist world, it is to the companies with the scale and expertise of global operation that these countries are turning to fill the vacuum left by the withdrawal of state control. This, in turn, must result in a further intensifying of the globalization process.

John Harvey-Jones, Chairman of ICI at the time, in his 1986 Dimbleby Lecture 'Does Industry Matter?' (Harvey-Jones, 1986), having referred to the high costs and risks of innovation in high-technology industry, went on to say:

> Thus, the UK has an absolute and inescapable need for international companies.... UK companies which are holding their own in fast-growing high-technology sectors do so because they are both large and international. On the other side of the coin some of the more successful companies in the UK are foreign owned. In 1981, for example, the hundred largest foreign firms in the UK accounted for 60 per cent of the total investment in the UK and were an important source of employment and taxes. They pay taxes just like the British-owned companies do and connect us into the fruits of their own research and development from their national bases. In all these ways we gain from internationalism and we should do all we can to encourage it – but equally, we need to have our own businesses operating overseas and as many of them as possible.

This compulsion to globalize on the part of industry is not matched by the attitudes of governments who find themselves both encouraging foreign investment and concerned about foreign domination. Industry, it must be said, is undeterred by the confusion and incoherence that it must face and overcome in developing a global status. Gladwin and Walter (1980), in their comprehensive study of the problems industry faces, open with the statement:

This book is about conflict. Not the textbook competition of the marketplace where the rules of the game are understood and concepts of winning and losing are part of the intellectual baggage all of us carry around. Not even the messier conflict that increasingly preoccupies managers in dealing outside the market with government and pressure groups solely at the national level. At least here the political and social elements giving rise to conflict are part of a more or less coherent national decision system that managers can comprehend and, more importantly, influence. Our concern is with conflict of a different sort, a veritable managerial nightmare where the rules of the game are often incoherent, ambiguous, redundant and ever changing. Where an attempt to resolve conflict in the here and now often leads to even greater conflict in another place or another time. Where managers are often trapped in conflict not of their own making. And where 'good' and 'bad' solutions are often indistinguishable at the time managerial decisions are made.

Entitled *Multinationals Under Fire – Lessons in Conflict Management*, every aspect of the problems faced is exposed. Bribery and corruption, human rights, terrorism, labour participation in management, ecology, political boycotts, tax avoidance and technology transfer are, as they say, 'a few' of the many dimensions they consider in a long and heavily documented study.

Despite all this, industry has accepted the global challenge. We now turn to the challenges posed to governments.

The challenge to sovereignty

The fear of foreign intervention, often unfounded, is always close to the surface in every society and no attempt to come to terms with the global challenge of innovation can afford to disregard its political power. Many of the criticisms of multinational enterprises – such as bad industrial relations, deceptive advertising, predatory pricing and monopoly – apply with equal if not greater force to wholly national enterprises. Nevertheless, they become much more important in popular judgement when used as examples of possible abuse of multinational industrial power.

These phobias – for they are often much more potential than proven threats – can be grouped under three headings:

1 *Politico-social* – the use of supra-national power to influence governments in a way not open to national industries to secure benefits to the company.
2 *Economic* – the movement of capital across frontiers outside the control of the governments involved. On the one hand this may be perceived

as damaging to the parent or host country in which capital investment is lost and in the recipient country as an aggresive investment aimed at 'knocking out' weaker national enterprises who have no access to investment on a comparable scale.

3 *Technological* – as in the case of capital, the transfer of technology across frontiers can be viewed as dangerous to the losing country and inadequate or inappropriate in the receiving country. It reaches its most paradoxical manifestation when the multinational does establish a research component of its business in a foreign country and is then accused of robbing national industries of that nation's technological talent!

The traditional power of sovereign states in respect of their economies, and the industries which form an essential part of them, has evolved to deal with nation-centred companies. Since industry, both in its wealth-creating and social roles, is perceived as a major factor in the nation's life, regulation of – if not intervention in – industrial affairs has been accepted by governments of all political persuasions as an essential part of their remit. How are they to answer to their electors when the globalized multinational enterprise becomes a major factor in the national life?

Furthermore, in a sense this new problem is a creature of a government's own actions. By withdrawing as an unsuccessful financier and manager in the macroinnovative field, it has left industry to find an alternative with sufficient financial strength to replace that of the state – and sufficient market strength to make commercial sense of the costs and risks involved. Either the state accepts the unacceptable – that it is powerless to influence the behaviour of a vital element in its economy – or it seeks international accords that match the supra-national power of industry with supra-national power of governments acting in concert.

This is the challenge – let us now examine some of the matters, currently in disarray, that cry out for attention through international understandings.

Competition policy, mergers and acquisitions

Concern at the preservation of competition is at the heart of many national regulatory practices. The United States, in a determined attempt to introduce objectivity, has developed a mathematical measure of market dominance to trigger off an investigation of a change which might endanger competition. The so-called Herfundahl-Harschmann Concentration Index (HHI) takes the sum of the squares of the market shares (measured in per cent) of all significant participants and compares the value before and after the proposed change. If the situation before had an

HHI between 1000 and 1800, then a change raising this by 100 points or more warrants an intervention by the anti-trust division of the Department of Justice. If the HHI is over 1800 an increase of fifty points is all that is required.

Into this sophisticated national system there now enters the transnational enterprise. In the case of US-based companies, the argument that a merger will make the combined firm more competitive against powerful foreign rivals is said to be becoming a 'fashion of the day'. However, it is also being applied in full measure to any transaction between a European and a US firm that may eliminate competition between them in the US. Furthermore, a merger between two European firms, both of which compete within the US and at least one of which owns US assets, is subject to the same US anti-trust laws as two US-based competitors. The consequences in terms of cost and delay before all the necessary documentation and hearings have resulted in a ruling are themselves quite formidable. It is not surprising that issuance of the preliminary injunction that starts this long process usually leads to the deal being abandoned.

The definition of the market in question, who the participants are, what the entry barriers to a new participant are, if abuse of market power occurs and the question of the benefits to the consumer of the economy of scale achieved still leave a large degree of arbitrariness in this attempt to introduce an objective criterion.

The UK has a Monopolies and Mergers Commission (MMC) to whom the Government or the Office of Fair Trading can refer any takeover or merger. Being concerned solely with the impact on competition in the UK – and the effect on prices in the UK that excessive market domination could bring – it is more appropriately employed at the relatively small industrial level. In the GEC references in Chapter 3, we have seen that a GEC/Plessey takeover was blocked on grounds of competition in the UK, whereas a Siemens/GEC combined proposal – introducing a foreign competitor into the UK scene – was given qualified approval. By this reasoning, takeovers of British companies by foreign macroindustrial enterprises are much less likely to be questioned than if they were by another British company. In the macroindustrial scene some critics see this as ensuring British industry is prevented from achieving critical mass on a UK scale and so is ripe for takeover by its foreign competitors. Fortunately, the Monopolies and Mergers Commission only recommends, the final decision being with the Government – and a widely-drawn interpretation of 'national interest' provides an escape route.

Germany has a powerful Cartel Office to which all mergers with a combined turnover of DM 500m (a relatively small figure in our context) have to be referred. It was on such a referral in the 1970s that the UK company GKN was refused permission to acquire a German company in a similar automotive component business. MAN were recently not allowed to buy the diesel engine activities of Sulzer on the grounds that this would result in MAN being the only supplier of large two-stroke engines for ships in Germany! This strict application of national competition rules in an increasingly global economy is similar to the UK position. Nevertheless, when applied to the Daimler-Benz massive market-dominating takeover of Messerschmitt–Bölkow-Blohm, the German Government overruled the Cartel Office on criteria related to global, rather than national, competition.

France has recently created a 'Competition Council' but it has only just begun to have an effect on internal competition within its national boundaries. In high technology, France has most of its activities in state-controlled or, at least, state-recognized national champions. A foreign bid for a French company can, by law, be stopped by the Government with a matching French bid. The issue of anti-trust is thus almost irrelevant in the absence of a free market for companies and the strong central control of industry exercised by the French Government.

Japan, as one might expect, does not need to address the subject of foreign takeovers. Their culture strongly opposes them, except in extreme rescue circumstances, on grounds of employee relations and loss of face for the management. Despite Japan's strong competitive instincts in its domestic market, it has not found it necessary to establish any anti-trust machinery. So, while everything is theoretically open, in practice, as one commentator recently said, acquisitions are as 'rare as Sumo wrestlers in pin-stripe suits'!

Interlocking shareholding of a few large shareholders, the restriction of shareholders' rights of decision, the secrecy of the share register, all combine to make many French, German, Dutch and Italian companies 'bid-proof' without any government intervention. In Switzerland, shareholders' rights are almost non-existent and management are immune from any control. This situation, which came to the notice of the British public in the Nestlé takeover bid for Rowntree, provoked an outcry against the blatant inequality of opportunity of British companies to takeovers of Swiss companies.

This inequality – and a certain outburst of nationalism – has also provoked a reaction in the United States. The recent Exon-Florio amendment is aimed at strengthening the power of the President to intervene in foreign acquisitions. The Committee on Foreign Investment in the US (CFIUS), set up by a 1976 Act, can recommend to the President that an acquisition be blocked. Before the amendment, however, the President would have had to declare a national emergency; now the criterion is 'national security' and would have been 'national security, essential commerce and economic welfare' if Congress had had its way. Further examples are cited in Chapter 9.

In the UK, the only restriction has been the retention by the Government of a golden share (giving it the right of decision in regard to change of ownership) and the restriction on foreign shareholding in its privatization of state-owned companies. In the case of Jaguar, the golden share restriction was withdrawn – to the embarrassment of the Jaguar directors – at a late stage and Ford was able to acquire the company. In the case of Rolls-Royce and British Aerospace, the share remains, but approval for foreign shareholdings to increase from 15 to 29.5 per cent has recently been given under pressure from the European Commission.

In summary, the regulation of competition between globalized industries is in great disorder. If there was a case for national procedures for ensuring competition, then the lack of their equivalent for multinationals is a matter of concern. The current disparities between nations are extreme and make the choice of parent country for a multinational a source of strategic advantage in escaping anti-trust and cartel control, as in the Nestlé takeover bids.

The Commission of the European Community, as in many transnational problems, is making some progress in respect of its member nations and could well, as in other aspects of international regulation, provide a pilot model of the way this might be tackled. Much coordination between the Commission and member states has to be agreed and implemented.

Taxation

The multinational enterprise presents the various nations in which it operates with both the problem of how to tax it and the suspicion that it is manipulating its accounts to minimize the amount it pays. This suspicion arises from the arbitrariness (in theory at least) of the prices used by one subsidiary in transferring work to another in the same company. The answer to this in most countries is to demand that all transactions across national boundaries between subsidiaries should be on an 'arm's length' basis and the profits of each subsidiary for taxation purposes should be calculated on this basis (more on this subject – in Chapter 9).

The alternative view, of which the Franchise Tax Board of California is the prime exponent, is that the notion of an 'arm's length' price is an outdated fiction for companies transferring complex components for which there is no external market across frontiers to assembly lines elsewhere. Their solution, the so-called unitary taxation concept, is to take the enterprise as a whole and to assign to the various tax authorities an element of the total profit in proportion to the 'activity' in that authority's area. The definition of 'activity' used is based on the average of the ratios of a company's sales, of its real estate and of its payroll which is then ratioed to the worldwide totals of these items in the company's books.

Despite vigorous protestation by the UK and Japan, a number of states in the US are determined to continue with this method of assessment. The simplicity of this concept from the tax-collecting point of view could make it attractive to other countries such as Nigeria and India. However, from the multinational's point of view the calculations and analysis to demonstrate compliance are considerable and it is reported that corporations (such as BP and BAT), who would pay less tax on this basis, have joined the lobby against it. It therefore remains an unresolved issue of considerable importance in the multinational scene.

The same arbitrariness arises in relation to the balance sheet of a wholly-owned subsidiary. National taxation has, traditionally, recognized two forms of financing a company – debt and equity – and allowed the interest on the former to be 'tax deductible' and the 'profit before tax', the reward of holders of the equity (the shareholders), to be taxable – both when it is distributed to shareholders and, through withholding tax, when it is retained in the enterprise. There are thus potential tax advantages if the balance sheet of the subsidiary is high in debt (to the parent) and low in equity. The response of the tax authorities, when they consider such 'gearing' to be unreasonably high, is arbitrarily and therefore contentiously to redefine some of the debt as equity and so treat part of the interest charge as if it were a dividend.

Differences between nations also exist in respect of the definition of residency – the criteria by which the nationality of the company for taxation purposes is determined – and these can create anomalies which become matters of contention. Sometimes the test is the place of registration, sometimes the place in which management and control is exercised. A loophole recently closed in both the USA and UK in this area allowed a company to establish dual residence in both countries and so to claim interest tax relief on the same borrowings twice over.

Tax avoidance through the ingenious use of countries with negligible or exceptionally low tax rates, particularly as the centres for the financial and treasury activities of a macroindustrial enterprise, is a long-established activity against which governments are continuously taking countermeasures. The Netherlands Antilles, Cayman Islands, Liechtenstein, Monaco and even Switzerland feature in the list of countries which play this role in the macroindustrial scene. They feature, openly, as devices to charge other highly-taxed subsidiaries management fees which shift profits out of highly-taxed countries into the tax haven. A recent ruling of the UK House of Lords Judicial Committee is leading the way in an attack on this practice. This established a doctrine which allows the taxing authority to ignore any step in a corporate activity inserted solely for the avoidance of tax. With the onus of proof on the company that it has a legitimacy, apart from tax avoidance, this can be a powerful weapon in the hands of the authorities.

All this is, of course, part of the wider problem of national taxation in a global context. Personal taxation and the taxes on consumption such as value-added tax, create discrepancies which an increasingly mobile labour force and flow of goods may be expected to exploit. The market for advice on how to choose one's residence for taxation purposes, once restricted to the very rich, now embraces the ever-growing number of expatriate workers in multinational industries. The differences in VAT between neighbouring countries – the cross-border traffic between Eire and Ulster is in this respect just one of many examples – illustrate the anomalies that arise from incoherent national attitudes to commodity taxes.

Regulation of trade – the General Agreement on Tariffs and Trade (GATT)

Since its foundation in 1947, the General Agreement on Tariffs and Trade (GATT) has been the subject of seven rounds of wide-ranging multinational negotiations aimed at coordinating and regulating national practices in regard to trade. In this time, global trade has grown enormously and with it the importance of agreements that facilitate its conduct in as mutually satisfactory a manner as possible.

The rules of international trade embodied in the GATT are not intended to be mere commercial conventions but are binding obligations on all its ninety or so members. There is, therefore, in place an organization capable of dealing with this aspect of the globalization of industry. While its achievements have been considerable, and it is not surprising that much remains that is unsatisfactory, it exists under the constant threat of unilateral or bilateral behaviour by its members.

Theoretically 'free trade' is the rule, but as the old-fashioned tariff barriers have come down, so new more subtle non-tariff barriers have gone up. Japan is accused of maintaining elaborate non-tariff barriers against imports from the US and Western Europe. Her European customers are in turn accused of demanding excessive local content before a Japanese manufacturing unit in Europe can claim its products are European. Tariffs come back into play to penalize a country whose export prices are judged to be so low as to constitute 'dumping'.

Into this crossfire of charge and counter-charge comes the mighty US with a burning conviction that a major part of the deficit in world trade she is currently experiencing is because other countries are not 'playing fair'. The result has been some controversial anti-dumping measures and the passing by the US legislature in 1988 of an aggressive Trade Act. Under the label 'Super 301', the Act requires the US Trade Representative (at the time of writing Mrs Carla Hills) to pursue the 'elimination of priority unfair practices' which slow or deny market access to US companies. Negotiations with these countries are required to proceed over a three-year period after which, if progress is not made, retaliatory measures must, under US law, be taken.

The burgeoning of Japanese manufacturing units around the world, the products of which are claimed to have the host countries' nationality, and not Japanese, is a continuing source of contention. By this means, Japanese cars made in the UK become European and Japanese cars made in the United States American – defeating whatever agreements have been made to limit the number of Japanese cars entering the European market. Assembly or 'screwdriver' plants, as they are called, are not enough and percentages of the total manufacturing cost of 40 per cent, 60 per cent, and even 80 per cent, are demanded to be incurred in the specific locality before nationality of manufacture can be claimed. Since it is of the essence of this industry that it is global in its make or buy decisions, this concept of strict locality of manufacture bristles with contentious and counter-productive issues.

Dumping leads to similar contentious issues when applied to highly innovative products, where the bare prime cost may be minute compared with the variety of 'full costs' which can be calculated when investment in innovation is amortized. The case of alleged dumping of D-Ram microchips by Japan in the USA exemplified another difficulty. When an anti-dumping tariff was put on their importation, there were as many complaints from the US importers and users of these products (who benefited from their low price) as there had been previously from the US producers whose prices had been undercut.

Industrial property rights, the protection of rights under national patent and copyright legislation, is far from satisfactory on a global scale. In attacking this problem from the United States point of view, the authorities there have estimated that the losses inflicted on the US economy by counterfeiters, patent pirates and infringers amounted to some 40 billion dollars in 1986.

This was, in part, due to the inadequacy of legislation and enforcement in the newly industrialized countries but also to deliberate government policy in such countries as India and the South and Central American countries. Particularly in respect of pharmaceutical products, where the welfare of an essentially poverty-level society is at stake, they are not willing to pay the price of the wealthy nations' approach to patents. In India, the only type of patent it is prepared to recognize in such circumstances is restricted to the process of manufacture and to a period of seven years. This enables foreign drugs to be copied, manufacturing them by a slightly different process, and so avoiding the massive mark-up required to pay for the R&D involved.

In all this GATT, like the United Nations in its field, is attempting to achieve solutions in a single step from a wide diversity of national interests to a global answer. It is not surprising that the difficulties are immense and the progress slow. It has been suggested that, in some respects, regional understandings – say covering South American, African and South East Asian nations – would be easier to achieve and then build up to a global agreement. The EC is such a region and could, if it avoids its own protectionist proclivities, take a step in this direction.

However, at the heart of the matter is the concept of sovereignty that nations bring to the negotiating table and their willingness to establish and accept procedures for the enforcement of agreements achieved. Weakness or failure to enforce compliance is very damaging. It is from this that, in frustration, nations resort to unilateral measures, reach bilateral agreements disregarding wider considerations and, in the longer term, create more problems than they solve. The US–Japan semiconductor agreement, the US Trade Act with its reprisal implications and the EC's 'Japan-bashing' proposals are all examples of such actions.

It is this difficult but vital concept of the globalization of sovereignty, already being put to the test in the approach to 1992 of the member states of the European Community, that is both the challenge and the key to the future of sovereign states in a world community more close economically than a single nation 100 years ago.

The globalization of sovereignty

The case for addressing the subject of the globalization of sovereignty extends far beyond the problems of multinational enterprise with which we are dealing here. The powers of destruction through nuclear warfare, through the abuse of the environment and the exploitation of natural resources all demand a global response in which the narrow interests of nations must be subjugated to the greater good of the planet as a whole. Innovation, with is own powers of abuse as well as usefulness, just adds further weight to the need to rethink this subject. This has been done in a most stimulating way by Lionel Stoleru (1987) in his book *L'Ambition Internationale* and by Christopher Layton (1986) in *Europe and the Global Crisis*.

In his introduction, Stoleru puts it most graphically by saying that we have the equivalent of a Copernican revolution in which we must recognize that (translated from the French):

national economies turn round the international economy like the earth turns round the sun, and not the reverse.... Abandon the illusion of strictly national policies modified by international constraints and face today's reality. International policies modified by national constraints.

If a distinguished French professor at the École Polytechnique, who has held high positions in the French Government, sees internationalism this way, it has most certainly arrived as a serious consideration!

The consequences of this axiom are complex and controversial. He makes it easier to accept by drawing on the analogous position of the individual in a national regime – circumscribed by the laws and disciplines of society but still much 'freer' than he would be without them. His overall plea, however, is that (translated from the French) 'it is now time to look at our problems through the other end of the telescope – the international end – because it is only in this way that we will find solutions to them.'

As we have already pointed out, the task of creating a single economic community in Europe opens up many, if not all, of the problems of the multinational scene. The spread of interest from Greece to Germany, from Portugal to the United Kingdom – and the added complexity of the emergent Eastern European democracies – provide an aggregation of problems whose solution could be a model from which future global understandings could be derived.

Christopher Layton has long-standing experience of working in the European Commission. He is an Honorary Director General and a former chef de cabinet of the Commissioner responsible for industrial affairs. In his book (1986) which he states is an essay, 'inspired by the belief that the time has come for a united Europe to take up responsibilities for working for a more united world', he identifies sharing of sovereignty as a central issue.

His plea, like Soleru's, is that there is no other way forward and that the debate should not be on how to retain yesterday's concepts of sovereignty but how to reproduce on a global scale the attitudes which, in their day, created unity within nations. In this the benefits – and dangers – of the growth of multinational innovative enterprise have a part to play. In the next two chapters, we examine in more detail the relationships between multinational enterprises and sovereign states and discuss ways in which conflicts can be resolved.

References

Gladwin, T. N. and Walter, I. (1980), *Multinationals Under Fire – Lessons in Conflict Management*, Wiley.

Harvey-Jones, J. (1986), 'Does Industry Matter?', the 1986 BBC Dimbleby Lecture. *The Listener*, 10 April.
Layton, C. (1986), *Europe and the Global Crisis*, Federal Trust for Education and Research, London.
Porter, M. (ed.) (1986), *Competition in Global Industries*, Harvard Business School Press, Boston.
Stoleru, L. (1987), *L'Ambition Internationale*, Editions du Seuil, Paris.

9 Multinational enterprises and sovereign states

- **Definitions**
- **Growth of MNEs**
- **The case of the automobile industry**
- **Two views on MNEs**
- **The mismatch between distribution of resources and markets**
- **The move of an MNE into a host country**
- **The positive role of the MNE**
- **Criticisms in the home country**
- **Criticisms in the host country**
- **International conflicts**

Definitions

Chapter 8 presented a broad canvas of the global and political scene in which multinational enterprises have to operate. In this chapter we continue the discussion and consider in greater detail some of the main issues involved.

The first point to note is that there is no universally accepted definition of an MNE. MNE stands for 'multinational enterprise', a term which many writers prefer to that of MNC, namely 'multinational corporation', since some MNEs are government controlled and are not corporations in the conventional sense. UN agencies prefer the term 'transnational', as some enterprises are not located in enough countries to make them truly multinational in the strict meaning of the word.

Madden (1977), in his introduction to a set of papers on multinational corporations, published a few years ago, says:

> Because they are successful and powerful, multinational corporations are the cause of much controversy. But even critics admit that internationalization of production must continue. By internationalizing operations, only the multinationals, of all business organizations, have effectively reacted to the prime world economic reality: the unequal distribution of resources human and physical. By concentrating on factors of production and not upon political

boundaries, they have been able to organize, produce and market on a worldwide scale, effectively mobilizing resources as no comparable economic unit operating from a national perspective can.

Madden thus encapsulates the arguments extolling the immense contribution of MNEs to world trade and hints at the criticisms levelled at their operations. As indicated in Chapter 8, neither the praise nor the criticisms have abated over the years; on the contrary, with the inevitable increasing interdependence of national economies, and the increasing power of the multinationals, the debate about their roles and responsibilities has been put into sharp focus.

MNEs come in different shapes and sizes, they are involved in a different range and scope of operations, they have different organization structures and they exercise different modes of control. Consequently, there is no reliable estimate of the number of MNEs worldwide. The only characteristic that they have in common, as their name implies, is that they have manufacturing plants in more than one country, but if that were their sole defining criterion, then the number of MNEs would be vast, since any investment abroad can easily fall into this category. Consequently, it has been suggested that merely investing abroad is not enough as a qualification for an MNE accolade, but that owning or controlling plants abroad is what matters, although what is meant by 'owning' or 'controlling' is open to varying interpretations. To some a minimum equity stake of 50 per cent is the crucial test, though others have suggested 30, 25 and even 10 per cent (Edelstein, 1982; US Dept of Commerce, 1970).

In the absence of a widely accepted definition and adequate statistics, it is also impossible to trace the growth in the number of MNEs over the years, in spite of the many books written on the history of MNEs and their development in general (see, for example, Hertner and Jones, 1986), or books devoted to the history of particular enterprises (such as DeLamarter, 1986; US Dept of Commerce, 1960). It is evident, though, that MNEs have been around for a long time. Some point to the British, Dutch and French East India Companies and the Hudson Bay Company in the seventeenth and eighteenth centuries as early examples of MNEs, though in their early days these companies were mainly concerned with the mercantilistic objective of supplying raw materials to the industries in their home countries, rather than with manufacturing abroad. By the Second World War it is estimated that the number of MNEs already amounted to several thousand (Hertner and Jones, 1986), and since then they have experienced a phenomenal growth, both in size and in numbers. At the beginning of the century the MNEs were all European and US in origin, but in recent years new MNEs from other countries, such as Japan, Brazil and Korea, have joined the established MNE ranks.

According to the UN Centre on Transnational Corporation's estimate (Stopford and Dunning, 1983), direct investments abroad in 1960 amounted to $67bn, of which $66bn came from developed countries; but the total level of investment mushroomed to $512bn in 1980, of which developed countries were responsible for about $500bn, or some 98 per cent of the total, with over a half coming from US corporations and about 15 per cent from the UK. Perhaps the most significant figures of the statistics for direct investment abroad relate to Germany where the figures rose from $1bn in 1960 to $38bn in 1980, and for Japan where the figures were $0.5bn and $37bn respectively, indicating the rapid expansionist plans of MNEs in these countries.

Growth of MNEs

More to the point, however, is the concentration of power that has become quite evident. For example, it has been estimated by the statistics of the *World Directory of Multinational Enterprises* that less than 5 per cent of MNEs account for more than 80 per cent of all foreign affiliates, that MNEs perform some 25 per cent of the world's total production of goods and services (Madden, 1977), and, as one would expect, that many of the companies involved are among the top 500 US and non-US corporations listed in *Fortune* magazine (Stopford and Dunning, 1983). The sales revenues of the top ten in 1981 were as shown in Table 9.1 (in billions $).

Table 9.1 *Sales revenue of the top ten corporations 1981*

Corporation	Sales revenue (billion $)
Exxon	113
Royal Dutch/Shell	82
Mobil	69
General Motors	63
Texaco	58
BP	52
Standard (California)	45
Ford	38
Standard (Indiana)	31
ENI	31

These are staggering figures, and it has often been remarked that the fact that they exceed, by a wide margin, the total budgets of many governments in developing countries or LDCs (less developed countries), immediately highlights the immense power of these MNEs and the resulting suspicion with which their activities are regarded in many quarters.

It is interesting to note that the massive concentration of the operations of the top ten are in the oil and automative sectors, which are quite different in character from each other. The concentration in the oil industry indicates the international nature of this business, with exploration and extraction taking place in countries rich in oil reserves, while marketing and distribution operations aim at worldwide markets. The growth of sizable operators in the automobile industry stems from the effect of economies of scale, where capital costs for new manufacturing facilities and for launching new models have become so huge, that the necessary sales volumes can only be achieved by a global manufacturing and distribution strategy. Thus, on the whole, the major suppliers of crude oil are not the major consumers of derivative petroleum products, while in the case of automobiles the reverse was historically true, with major manufacturers concentrating first on supplying their home markets before turning their attention to investment in manufacturing facilities abroad.

While oil and automobiles continue to dominate the official statistics, other industries are beginning to exhibit similar patterns of international growth and penetration. Hard on the heels of the top ten in Table 9.1 come industries such as chemicals, mining, pharmaceuticals, computers and telecommunications. Chemicals (particularly the petrochemical industry) and mining are similar to oil, in their dependence on access to natural resources of minerals located in countries where the demand for the final product is low, compared with that of the highly developed industrialized countries. At least in principle, the owners of the natural resources and basic commodities can bargain with the MNEs from a relative position of strength as to how the benefits of exploration and extraction can be shared, and there are indications of mineral exporting countries becoming more sophisticated and more demanding in their negotiations.

In contrast, the future development of pharmaceuticals, computers and telecommunications depends far less on the availability of scarce mineral resources and much more on advanced technology and the fruits of R&TD, all of which are under the direct control of the MNEs in industrialized nations. The nature of their dependence on underdeveloped or developing countries is confined to dependence on their markets, with the powers of the sovereign states constrained to determining the conditions under which the MNEs would be allowed to trade and operate in these markets. These powers should not be underestimated, particularly when competitors among MNEs in the same industrial sector vie with each other to secure access to potentially promising markets, but the negotiation framework is not as menacing as when the supply of crucial raw materials is at stake.

Studies of the history of MNEs often highlight two phenomena: first, the inexorable growth in their size over the last forty years, leading to an ever-increasing share of the market under their direct control, and second (as a corollary of the first) the decline in the number of competitors in any sector in which MNEs dominate. One prediction, for example, is that by the year 2000 the major part of world trade would be controlled by some 300 corporations and that the number of major MNEs will settle at around that number. In fact, the evidence for these assertions is at best patchy. Many MNEs have experienced phenomenal growth (as shown by the example in Table 9.1), others have floundered in spite of their rosy prospects. An examination of the *Fortune* top 500 or the *Times* top 1000 reveals that the relative position of many of those occupying prominent positions in the league tables changes constantly, and many a household name has disappeared from the lists altogether and given way to newcomers. Being among the leaders gives a company no immunity and no guarantee that it will stay there.

The case of the automobile industry

Such complete changes of fortune are evident even in a mature industry, such as automobiles (briefly discussed in Chapter 3), giving rise to a very large number of entrepreneurs at the beginning of the century, leading to mergers, takeovers and consolidation. Many well-known names have completely disappeared, or have been submerged under larger corporate umbrellas, and the general impression is that the number of independent manufacturers has been drastically reduced over the years and that production has become increasingly concentrated in the hands of a small number of corporations.

Judged over a long period this development is indeed discernible, but the fear that the trend will continue and will eventually wipe out all the small companies in the industry, or debar the way to newcomers, is unfounded, as we can see from Table 9.2, which shows the number of cars produced in Britain in 1988 compared with 1981. First, it is interesting to observe that, contrary to common belief, the number of manufacturers has not dropped during that time and we note that while ten manufacturers were listed for 1981, the number rose to fourteen by 1988. The top two manufacturers retained their positions in the league table (Rover and Ford), while the relative positions of the next two (Peugeot/Talbot and

Table 9.2 Car production in Great Britain (number of units)

	1981	1988
Aston Martin	157	214
Carbodies	–	2332
De Lorean	7409	–
Ford	342,171	375,542
Jaguar/Daimler	(incl in Rover)	51,939
Lotus	345	1336
MCW	–	909
Nissan	–	56,541
Panther	–	229
Peugeot/Talbot	117,439	82,326
Reliant	89	201
Rolls-Royce	3087	2968
Rover Group	413,440	474,687
TVR	167	701
Vauxhall	69,932	176,489

Source: Society of Motor Manufacturers and Traders, 1989

Vauxhall) was reversed, with Nissan (a newcomer in the 1980s) hard on their heels. Second, the top four manufacturers accounted for 98.8 per cent of the total in 1981, but their combined share fell to 90.4 per cent in 1988, largely because of the impact of Nissan.

On a global level, however, the distribution between the major trading blocs has remained remarkably stable during the 1980s, as shown in Table 9.3. The percentage of car production of the world total attributed to the three trading blocs (the EC, North America and Japan) was 85.4 per cent in 1982 and remained virtually the same in 1988. In spite of the Japanese imports, the percentage of production in the EC also remained virtually the same (rising from 37.3 to 37.5 per cent), while the percentage of North America rose with a corresponding fall of the percentage of production in Japan. What these figures do not reveal, however, is the substantial investment of Japan in production facilities within the EC and the US in recent years, so that cars manufactured in these areas appear under the EC and North American banners.

It is further interesting to note that the top ten car companies in the world, each producing more than one million cars per annum, belong to the three trading blocs. Three are American, three are Japanese and four European (see Table 9.4), though these identities refer more to the origins of these companies rather than to their centres of operations, since they all have manufacturing plants in several countries and are not confined to a single trading bloc. Between them these ten companies produce about 58 per cent of the world output of passenger cars.

Table 9.3 World production of cars

	1982		1988	
	'000	%	'000	%
Economic Community	9,985	37.3	12,624	37.5
US and Canada	5,881	22.1	8,136	24.0
Japan	6,882	25.8	8,198	24.2
The three blocs as % of total	–	85.4	–	85.5
Total of 25 countries	26,638		33,870	

Source: Society of Motor Manufacturers and Traders, 1989

Table 9.4 The top ten car producers in 1986

Company	Number of cars ('000)
General Motors	4,316
Toyota	2,684
Volkswagen/Audi	1,774
Nissan	1,769
Ford	1,764
Peugeot/Citroën	1,468
Fiat	1,463
Renault	1,305
Chrysler	1,298
Honda	1,025

Source: Society of Motor Manufacturers and Traders, 1989

The dramatic increase in exports of Japanese cars is characterized by the figures in Table 9.5, with North America being a prime target, as indicated in Table 9.6. The imbalance of trade in cars between Japan and the US is also shown in Table 9.6. Some of this blatant imbalance has been attributed to the ability of the Japanese to adapt their designs and prices to suit the needs of the American market, but as all foreign car manufacturers (and not just the Americans) have failed to establish a foothold in the Japanese market, restrictive practices on the part of the Government and motor traders in Japan have been blamed for the growth of this distinctly one-sided pattern of trade.

The charge of obstructing car imports is not confined to Japan. The existence of restrictive practices, official or otherwise, can be legitimately

Table 9.5 Growth of exports of two major Japanese car producers

	Honda ('000)	Toyota ('000)
1960	–	2
1970	19	346
1980	651	1,149
1986	698	1,210

Source: Society of Motor Manufacturers and Traders, 1989

inferred from the figures in Table 9.7, which shows the level of Japanese car imports into various countries in 1976 and 1988 respectively. In all the countries listed (except Portugal) the figures rose during the twelve year period. In countries without an indigenous motor industry, such as Australia, Denmark, Ireland, Norway and New Zealand, the figures rose dramatically (in Australia almost half the cars are imported from Japan, and in New Zealand the level is about 60 per cent). In countries with their own motor industries the growth of Japanese imports has been much slower, either as a result of agreements with Japan on import quotas, or through the awakening of the competitive spirit of local manufacturers, or both. But there are three countries listed in Table 9.7 which managed to resist the Japanese onslaught, namely France and Italy, who have been determined to protect their long-established car producers, and Spain, who is a relative newcomer to car manufacture and assembly and is keen to ensure that its burgeoning industry is not washed away in the flood of Japanese imports. Although all three countries are subject to the same rules and regulations that govern the other members of the European Community, Table 9.7 shows that forces other than the free market are active to keep predators at bay.

Table 9.6 Car exports

	Japan to US ('000)	Japan to Canada ('000)	US to Japan ('000)
1960	1	–	3
1970	324	65	5
1980	1,819	158	10
1986	2,348	222	3

Table 9.7 *Percentage of Japanese imported cars*

	1976 %	1988 %
Australia	30.0	44.9
Austria	6.0	26.9
Belgium	18.0	20.8
Denmark	16.5	35.1
Finland	21.9	40.4
France	2.7	2.9
Germany	1.9	15.0
Greece	15.1	28.6
Holland	16.8	24.4
Ireland	12.7	43.4
Italy	0.01	0.5
New Zealand	26.5	60.2
Norway	28.3	35.0
Portugal	16.1	9.8
Spain	—	0.6
Sweden	8.2	20.9
Switzerland	5.8	26.7
UK	9.4	11.2
US	9.3	20.8

Two views on MNEs

The growth of MNEs and their increasing influence on world trade, raise questions about their perceived impact and eventual role, as their complex love–hate relationships with sovereign states continue to unfold. We shall turn to postulate about the future in Chapter 10, but first we need to consider two contrasting views of the role and activities of MNEs. One view is that MNEs have made an immense positive contribution to the economic progress of all nations, and in particular of LDCs (less developed countries). The second is that the very phenomenon of the MNEs is a manifestation of the pursuit of power, by individuals and by corporations, designed to dominate markets through the ruthless exploitation of weak governments and the manipulation of national economies for their own ends. Both views are founded on legitimate arguments with convincing supportive evidence. But like many other issues in the economic and political arena, the picture is seldom either black or white. Let us examine some of the arguments in more detail.

The mismatch between distribution of resources and markets

The positive view of the MNE is based on the proposition that all MNEs are a manifestation of the needs and aspirations of people throughout the world. There are vast numbers of people, particularly in LDCs, who have basic needs for food, clothing, shelter and education. When some of these basic needs are met, the populations in all countries continue to clamour for an improved standard of living, for available and affordable consumer goods, for improved transport systems, for better health and social services. Even in highly industrialized countries, where the basic needs for a large proportion of the population have been met, people's appetites for improved living conditions and for increased consumption of goods and services have not abated (in fact, it appears to be higher per capita than in LDCs).

The enormity of the problem is put into sharp focus when we also remember that the world population continues to grow. Two hundred years ago the world population stood at about 750 million with an income per capita of $200; today the population is over 4 billion with an income per capita of $900, and it is estimated that the population will reach some 6 billion shortly after the year 2000. It becomes evident, therefore, that even the current average needs have put an enormous strain on the world's physical resources, and that this strain is expected to grow at an astounding rate.

What is equally evident is that the resources are not distributed in the world in proportion to population densities, or in proximity to centres of demand. Few industrialized countries are self-contained, in the sense that they have sufficient mineral and fuel resources to sustain an indigenous industry and to meet all the material needs of their own populations (even the US, which for many years was thought to be self-sufficient for most physical resources, can no longer claim to be in that position). It follows, therefore, that for the needs of the population to be met, raw materials have to be transported from their countries of origin to industrial centres which have the facilities and skills to convert the raw materials into finished goods, and these goods have to be transported again to market destinations to satisfy local demand.

Four types of centres can, therefore, be identified (see Figure 9.1):

- *Centres of origin* – several countries of origin are often involved to provide the physical material inputs needed for any given final product.
- *Centres of manufacture* – here, too, several countries may participate in the process of manufacturing parts and subassemblies to feed the final assembly line.

- *Centres of destination* – these represent the final markets, where the products are sold and where the sales revenue is generated.
- *Centres of planning and control* – to engage in R&TD, to design and plan the product range, its manufacture and distribution, to marshal the financial resources needed for the enterprise, and to control and coordinate the many activities required to convert the plans into reality, including the overall management of the enterprise.

It is because all these centres do not coincide geographically, and because their resources are not identical, that the interdependence of national economies has become inevitable. At the same time, mechanisms had to be devised to ensure that the diverse interests of mineral exporting countries, manufacturers, consumers and providers of capital and know-how can be interwoven to make it worthwhile for all concerned to collaborate. It has long been recognized by economists, politicians and businessmen that international trade is not a zero-sum game and that, if the right terms are struck, all the participants stand to gain from the outcome. Success of collaboration, therefore, depends on the skill with which the 'right terms' can be determined and by the effectiveness of management in implementation.

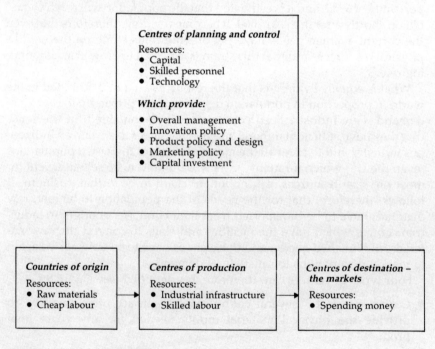

Figure 9.1 *Four centres of activity*

The geographical dispersion of the various centres shown in Figure 9.1 suggests that the physical transportation of raw materials and finished goods must play an important part in the total cost of the operation, particularly when heavy or bulky goods are involved. One way of reducing the effect of transportation costs is to arrange for some of the centres or locations to merge. For example, the manufacturing facility may be located either in the country of origin of the major raw materials, or in the country of final destination near the market. Apart from a reduction in transportation costs, such solutions can save in inventories and work-in-progress, involve shorter lines of administrative communications and can thereby contribute to improving the MNE's managerial and financial performance.

However, many technical considerations need to be addressed; for example, there may be several countries of origin and several market destinations, and some may not have the necessary resources or conditions to support large-scale manufacturing entities. In addition, the basic conflict between countries of origin and countries of destination remains, and consequently political issues need often to be resolved. There are, of course, many cases where the establishment of a manufacturing facility is not determined by the geographical locations of the sources of raw materials and the markets, but by other considerations, such as the availability of relatively cheap and productive labour and various direct and indirect government incentives in less developed and developing countries, adding, thereby, to the complexity of international economic and political relationships.

This is where the MNE has found an obvious opportunity to play a vital role:

> In a world of virulent nationalism, the multinational has offered a most effective way to use world human and natural resources. It has offered an adaptable way for people of different cultures, ideologies, and values to work together. It has offered an effective way of transferring a package of capital goods, management, marketing know-how, and technology from one country to another. It has excelled in modern management and in producing and marketing innovative goods and services (Madden, 1977).

The move of an MNE into a host country

Before we proceed to explore the relationships between an MNE and its home country on the one hand and the host country on the other, we need to review briefly the motives that encourage the MNE to set up a manufacturing and trading entity in a host country. There is a vast literature on this subject (see, for example, Caves, 1982 and the many references in Hertner and Jones, 1986 and Madden, 1977). In the context of this book, however, perhaps it is appropriate to highlight only four main motives.

The first is the desire to penetrate many markets and to achieve a large enough sales volume in order to generate the profits needed to become a global player with adequate resources to engage in macro developments, as discussed in the early chapters of this book.

The second, as intimated earlier in this chapter, emanates from Figure 9.1. Under the relentless pressures from competitors, the MNE is forced to pursue greater efficiency in order to achieve a reduction in unit costs. This objective raises the question of whether a manufacturing facility in an industrialized country cannot be replaced by one located in a suitable host country. A cost comparison, taking into account wage rates, availability of labour skills, industrial relations and the attitudes of trade unions, the possible savings in transportation costs, the lure of government incentives (in the shape of tax holidays, development grants, and many other inducements mentioned in Chapter 8), the industrial and professional infrastructure – all these factors play a part in a comprehensive evaluation to identify suitable host countries for the MNE to consider and target.

The third motive may be seen as a rationale for a strategic manoeuvre to diversify manufacturing facilities, in order to reduce the MNE's dependence on the home facility, so as to become less vulnerable to changes of government policies, to perceived whims of labour unions, and to fluctuations in international economic and market conditions. Multisourcing of components and multiple assembly lines in several countries have been developed as part of an overall strategy to allow the MNE to remain in control when disruptions (natural or man-made) occur and to be able to realign its resources and activities accordingly.

The fourth motive is to overcome trade barriers, usually in the form of protective tariffs, import quotas or other restrictions. The purpose of these barriers may be merely to protect local manufacturers from the flood of cheaper imports. In the case of LDCs this may well be understandable, particularly when their local products have to rely on relatively small sales volumes, which could be in danger of being wiped out in the absence of adequate protection. Alternatively, the local government may wish to discourage imports, which cause an outflow of hard currency and become a burden on the country's balance of payments. Protective tariffs, import quotas and other technical and administrative barriers are not confined to LDCs. As we see from Tables 9.6 and 9.7, there are good reasons to suspect that free imports of cars have been resisted vigorously in the case of Japan, France, Italy and Spain. By proposing to establish a manufacturing plant within the boundaries of a host country, particularly if it is a less developed country, the MNE can offer many advantages (some of which are enumerated below) and succeed in penetrating the local market with the active support of the host government. The concern of Japanese car manufacturers that the US and the EC may be increasingly tempted to erect trade barriers against car imports has been the main incentive for heavy investments in manufacturing plants within these trading blocs, and other industries have been studying these developments very closely.

The positive role of the MNE

From the foregoing discussion in this chapter, the case for the MNE may be summarized as follows:

1 The MNE acts as a facilitator, by bringing together human and physical resources in order to provide goods and services to meet human needs. It may be possible for the necessary framework to be created by a series of bilateral agreements between corporations and/or governments, but the MNE is a much more efficient agent for the purpose, initiating the whole process, undertaking the many ensuing activities, and performing the essential function of planning and control to ensure continuity throughout.
2 The MNE has the ability to put together a financial package (either from its own resources or through the backing of financial institutions) and take the risk that is inherent in any large-scale undertaking. Neither the host country (where the MNE wishes to set up an extraction or manufacturing plant) nor local companies are willing to take the risk, or are able to marshal the necessary resources on their own.
3 The MNE has the experience, technology and know-how to design, produce and distribute the product. By assuming responsibility for the whole process depicted in Figure 9.1, and by being prepared to underwrite (by itself or with others) the risk involved in a complete or partial vertical integration of the many activities involved, the probability of success is greatly enhanced. The mineral-exporting country, being a potential host country, may be tempted to undertake the manufacture of the product (assuming that it can raise the capital and has access to modern technology and design, either by investing in RTD&D (research and technological development and design) or by the use of licence agreements) in the hope of reaping large profits. But having the raw materials and even the necessary manufacturing facilities is not enough. The involvement of the MNE through the whole value-added chain in Figure 9.1 achieves the advantage of economies of integration and thereby greatly reduces the risk that otherwise would face the

mineral-exporting country. The economies of integration put together the advantages of location, international production, global distribution, and above all, ownership. The attributes of ownership involve technical and managerial know-how, RTD&D, patents, capital, and capabilities for overall planning and control, all of which can be offered by the MNE, as summarized in the top box in Figure 9.1.

4 Furthermore, the MNE has the resources to continue to invest in RTD&D without which the product would not be able to withstand long-term global competition. As argued again and again in this volume, RTD&D, and particularly macroinnovation, demand both highly-trained and motivated engineers and scientists, as well as ample financial resources, and these are more easily marshalled by large corporations. The MNE is, therefore, well placed to safeguard the future of the product in that respect.

5 By having to compete in the marketplace against other competent corporations, the MNE strives to be efficient in all its operations: in determining the range of products, in using resources efficiently in order to reduce costs, in controlling quality, and in ensuring good value for money for the customers. In this way the customers benefit, the suppliers of raw materials benefit (in securing continued demand for their materials), the workforce benefits (in securing employment) and the MNE and its financial backers benefit (in securing a good return on their investment). Furthermore, the threat of competition and possible new entrants into the market ensure that it is in the MNE's interest to continue to modernize and remain efficient.

6 The potential benefits to the host country are many:

- *Creating new employment* – many host countries suffer from unemployment and are keen to increase employment opportunities, which bring prosperity and an improvement in the standard of living.
- *Inflow of capital investment* – for plant and machinery, for start-up activities, for initial working capital, and for future expansion, all of which improve the balance of payments.
- *Tax revenue from employment and from corporation profit* – these can be substantial and help to finance economic and social programmes.
- *Imports substitution and creation of export opportunities* – both help to save and/or earn foreign currency and thereby improve the balance of payments of the host country.
- *Introducing foreign competition to local manufacturers* – to encourage higher industrial efficiency and productivity; there are many examples of local monopolies being shaken by investments from abroad (watches and office machinery in the UK and car tyres in France are notable instances).

- *Strengthening the industrial base* – to provide new opportunities for investors in related industrial operations.
- *Acquisition of modern technology and increasing the pool of trained personnel* – with direct and indirect benefits to other industries in the host country.
- *Improvement of the infrastructure* – particularly in LDCs the MNE can be instrumental (entirely from its own resources, or contributing to government expenditure) in establishing road and rail networks, urban development, new schools, training centres, libraries, hospitals, housing and social amenities.

Criticisms in the home country

Against all these benefits there is a long list of charges and criticisms levelled at the MNE. First, in the MNE's home country there are powerful groups (government departments, politicians, trade unions, and even industrialists) that resent the export of capital and jobs, arguing that in the absence of a foreign venture both capital and employment could be retained at home. These arguments become particularly poignant when the balance of payments comes under pressure and when unemployment rises. Under such conditions, foreign investment is usually seen as having an adverse effect on both counts.

The counter-argument of the MNE is that if it does not undertake a proposed foreign investment some other competitor will, and that as manufacturing in an LDC or an NDC (newly developed country, such as Hong Kong, Singapore, or Korea) is bound to be cheaper than at home (because of substantially lower labour costs and possibly other inducements), the loss of jobs at home would eventually occur anyway, as the customer switches to cheaper (or better-value) goods. As for the capital outflow, it is argued that it should be regarded as an investment which would lead, in due course, to inflows of funds in the form of profits, dividends, royalties and management fees.

Another serious complaint by the government and politicians of the home country concerns the loss of tax revenue when operations and profits move abroad and when an MNE engages in various manipulations to reduce its tax burden. In its defence, the MNE again contends that a proportion of the profits is bound to migrate to where the manufacturing activities are located, irrespective of whether these operations are controlled by the MNE or by a foreign competitor, so that as manufacturing capacity migrates, the tax revenue generated by these operations would be lost to the home country in any event.

Excuses to justify manipulations that result in tax avoidance are perhaps less convincing and tend to focus on the usual argument that the management of the MNE has a duty to look after the interests of its shareholders and that it therefore has to consider all available means, including convoluted financial transactions (provided they remain within the law), that would add to corporate financial strength and performance. Such an attitude does not exactly endear the MNE to the politicians and it is not surprising that some home governments are under pressure to curb the MNE's freedom by various means, including restrictions on the movement of capital. In Chapter 8 we referred to the move in California and in some other states to impose tax on the total earnings of the corporation, including those belonging to subsidiaries and affiliates abroad, and this is one example of how governments are understandably loath to lose tax revenues.

A further charge against MNEs operating in LDCs has been the widespread use of bribery to win contracts and other favours from local government officials and politicians. Several scandals have been publicized in the US and Europe with strong criticisms of MNEs aiding and abetting in the conduct of corrupt and illegal practices, which – if unchecked – could contaminate the commercial climate around the world and undermine business ethics in the industrialized countries as well. The MNEs' defence has always been that when in Rome we have to behave as the Romans do, and that if bribery is the norm in a particular country, then an MNE can abstain only at the real risk of losing business to a competitor. Public outcry and further legislation in some home countries seem to have curbed the worst practices, but these instances have been cited as convincing evidence that MNEs cannot be trusted with self-regulation.

Criticisms in the host country

As for the host country, here too many criticisms and complaints are encountered and some are summarized below:

- A collaborative agreement granted to an MNE to operate in a host country often precludes entry of new competitors into the market, at least for a while, in order to allow the MNE time to recoup some of its investment. This monopolistic or oligopolistic position enjoyed by the MNE can lead to price increases and to reduced customer choice.
- In spite of initial promises to import modern technology and knowhow, which it is hoped would eventually spread and benefit industry as a whole in the host country, MNEs tend to keep the most advanced technology, as well as all R&D, to themselves and grudgingly release

information only when they have to, or when there is a risk of competitors outbidding them when contracts with the host country have to be renegotiated.
- In their desire to make the most of their ventures in LDCs, MNEs tend to ignore the interests of the local community and are content to damage its environment, culture and social structure. Instances of pollution, destruction of the countryside, introduction of traffic congestion and growth of urban deprivation and crime all add up to a sad record of neglect by MNEs intent on short-term financial gains.
- MNEs are often accused of evading taxes in the host countries by arranging to pay inflated royalties and management charges, and in the main by fixing high transfer prices for imported materials, components and services. All these manipulations serve to reduce the declared profits in the host countries and result in hefty reductions in their tax revenues (these and allied issues were discussed in Chapter 8 and are considered further in Chapter 10).
- MNEs insist on being allowed to conduct their financial affairs and engage in foreign currency operations without interference of the host government, but large-scale currency transactions in LDCs can easily destabilize local currencies and thereby undermine their national economies. MNEs' financial operations can cause serious distortions of the capital structure in the host country, resulting in wide fluctuations and uncertainties that affect the whole industrial base. When demand for capital and credit facilities increases rapidly, interest rates may rise and add further pressures on the local economy.
- A large MNE operation in the host country can create a serious shortage of skilled technical labour and a strain on professional services (lawyers, accountants, educational institutions), as well as on the infrastructure (roads, electricity supply, telecommunications). An investment by the host country (sometimes aided by the participation of the MNE) to relieve these pressures would have a beneficial influence on the local economy in the long run, but scarcities in the short term (for example of skilled personnel) can have a serious adverse effect on the operations of other companies in the area, increase their costs, and contribute to an increase in inflation.
- In the main, criticism of MNEs in LDCs has centred on loss of sovereignty by the host government, in spite of the often declared policy of the MNE not to engage in local politics. The sincerity of such declarations is questioned when there is a threat of a friendly regime being replaced by one less favourable to the MNE's cause, and when the long-term future of the MNE in the host country is in doubt, so that the incentive for the MNE to intervene, overtly or otherwise, may become irresistible.
- Suspicions of the motives and activities of the MNE are often deepened

by inadequate published information about the MNE's operations. Commentators and analysts often have difficulties in disaggregating the cost and profit data for individual product lines, of ascertaining the rationale for pricing policies, of understanding the structure and effect of transfer prices, and of identifying capital movements. The lack of clarity of information on all these issues is sometimes blamed on accounting conventions, but there are two further reasons for the MNE management being reluctant to divulge more information than it has to. First, many cost and profit data are regarded as confidential, in that they may prove damaging if allowed to fall into the hands of a curious competitor or a hostile predator. Second, the suspicions of the local politicians of the motives of the MNE are mirrored by the suspicions that the MNE has of the motives of local politicians, particularly if a change of government could have far-reaching consequences to future operations and profitability.

International conflicts

So much for the grievances directed at the MNE by the host country, and these are superimposed on those held by the home country, as enumerated earlier. The complex network of the relationships implied in Figure 9.1, bearing in mind that each of the boxes may involve several countries, indicates that in addition to bilateral conflicts between an MNE and each of the governments involved, there is also a potential overt or covert conflict between the governments concerned. As Vernon puts it (Madden, 1977)

> When the US Government lays down regulations that oblige a US parent to withdraw funds from its British subsidiary in London, sterling may suffer while the dollar may be given a boost. When the Mexican Government directs the Mexican subsidiary of a US-owned automobile company to produce components for export to affiliates in Detroit and Sao Paulo, Mexican workers may be helped while US and Brazilian workers suffer. When the US instructs a US firm not to permit its Mexican subsidiary to export to Cuba, US policy may be helped but Mexican policy may be hurt. When the Saudi-Arabian Government forbids Aramco (a partnership of four US competitors) to ship oil to its parent's bunkering stations that serve the Seventh Fleet, the interests of the US are placed in jeopardy. In brief, as long as the units of a multinational enterprise are located in different countries, there is always the risk – and often the reality – that the interests of one nation will be served at the expense of another.

Such conflicts of interest, some economic and others political in nature, highlight the enormous political and economic pressures under which MNEs have to operate. They clearly cannot please all the players all the time. All the indications are that these pressures will grow, unless all the parties concerned begin to realize the need for an acceptable conduct of behaviour, both by governments and by MNEs, and this issue is discussed further in Chapter 10.

References

Caves, R. E. (1982), *Multinational Enterprise and Economic Analysis*, Cambridge University Press.
DeLamarter, R. T. (1986), *Big Blue: IBM's Use and Abuse of Power*, Macmillan.
Edelstein, M. (1982), *Overseas Investment in the Age of High Imperialism: the UK 1850–1914*, Methuen.
Hertner, P. and Jones, G. (eds) (1986), *Multinationals: Theory and History*, Gower.
Madden, C. H. (ed.) (1977), *The Case for the Multinational Corporation*, Praeger Publishers.
Society of Motor Manufacturers and Traders (1989), *Motor Industry of Great Britain 1989*.
Stopford, J. M. and Dunning, J. H. (1983), *Multinationals – Company Performance and Global Trends*, Macmillan.
US Dept of Commerce (1960), *US Investments in Foreign Countries*.
US Dept of Commerce (1970), *1971 Foreign Direct Investment Program*.

10 Managing the international trade game

- Five types of players
- Bilateral and multirelationships
- Goals of the five players
- The national interest and the MNE response
- The effect of trade unions
- International corporate networks
- Political issues
- Transfer pricing
- The role of joint ventures
- Code of conduct

In Chapter 9 we briefly discussed the four centres of activity involved in international trade, and the physical flow of materials and goods is shown schematically in Figure 9.1. In fact, the four centres may involve several countries concerned with the financing, extraction, manufacturing and marketing process, and all this results in a complex series of relationships and diverse interests, as we shall see below.

Five types of players

The intertwining operations of the following five primary players involved in the conduct of multinational business (which is later referred to as *the international trade game*) are described in a simple macromodel in Figure 10.1:

1 *The MNE* (multinational enterprise), which assembles the resources and provides the management for the execution and coordination of all the financing, RTD&D, procurement, manufacturing and distribution activities.
2 *The home country*, where the MNE is registered, where it generally has its head office and where it pays tax on its consolidated accounts.

Managing the international trade game 179

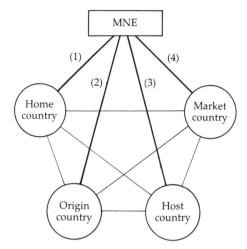

Figure 10.1 *A network of bilateral relationships*

3 *The origin country*, where the raw materials essential for the manufacture of the products come from.
4 *The host country*, where the MNE has manufacturing facilities, either for components and subassemblies, or for the final assembly operations.
5 *The destination, or market, country*, where the finished product is finally distributed and where the sales revenue is generated.

One of the five players is the MNE and the others are the various countries involved. As Figure 10.1 suggests, the geographical distribution and the different roles of these players give rise to a network of ten types of bilateral relationships, four of which concern the MNE's obligations to conform to the overall laws and regulations that apply to operations in each of the country-types, subject to any special agreements that govern the MNE's activities in the country. The four are:

1 The MNE's relationship with the home country mainly covers matters of taxation, adherence to the rules governing the export of capital and transfer of technology to foreign countries, compliance with constraints relating to politically-sensitive products, and generally behaving in a way that does not run counter to government foreign policy.
2 The MNE's relationship with the country of origin is primarily that of a purchaser of raw materials, but protection of the environment and concern for social conditions in the country are increasingly becoming important issues in which the MNE gets involved, either voluntarily or (more commonly) in response to pressure by the government and local politicians.
3 The role of the host country is discussed in Chapter 9 and shown as a vital ingredient in the value-added chain in Figure 9.1. As an employer,

the MNE has a responsibility to observe the laws and customs of the host country regarding employment, training, health and safety, payments of taxes, and protection of the environment. It also needs to contribute (in one way or another) to the economic and social infrastructure of the host country, such as in the areas of welfare, social and medical amenities, housing, education, transport and telecommunication systems.

4 The relationship of the MNE with the destination country concerns regulations governing imports, requirements relating to technical and safety specifications of products offered for sale, rules about trading practices and management of distribution channels.

At its simplest, the model in Figure 10.1 reduces to that of a single country, which encapsulates all the four centres of activity shown in Figure 9.1. It has all the necessary raw materials, it accommodates within its boundaries the production plants, and it constitutes the market for the final product. If the enterprise is registered in that country, then the roles of home, origin, host and market countries are all rolled into one. This may be called the 'singular' model. At a macro level, the relationship of the enterprise (which in this context does not operate as an MNE) with such a self-contained country seems relatively straightforward. However, at a micro level there are, within a single country, many interested parties which the enterprise has to deal with, such as the national government, local authorities, trade unions, professional and trade organizations, regulatory bodies, banks, shareholders, suppliers, distributors and customers.

This wide range of interested parties results in a formidable array of relationships, even in the singular case in which only one country is involved. When this array is superimposed on the model in Figure 10.1, with several countries taking part in the trading system, the underlying complexity of the MNE's planning and control process can be fully appreciated.

Figure 10.1, therefore, represents a very wide spectrum of circumstances and combinations, as the following three examples illustrate:

1 Departing from the singular model, the enterprise decides to reduce its production costs and establish manufacturing facilities in a foreign country, but it intends to import the finished product for sale in the home market. Two countries are then involved: the home country (which in this case is also the market country), and the host country.
2 Following from the previous case, the enterprise proceeds to sell some of its output in the host country (which now becomes a market country as well), in addition to continuing to supply the home market (which also serves as a market country). Here we still have only two countries involved, but each has a dual role.

3 If the raw materials are not available in the home country, the MNE may determine to establish its manufacturing facilities in close proximity to the source of materials and transport the finished goods for sale in the home country. Again, only two countries are involved, but here the origin and the host countries coincide, as do the home and market countries. If some of the final output is sold in the host country, then it performs a triple role (origin, host and market).

There are, of course, many other combinations that involve two or three countries. At its most elaborate, the macro model in Figure 10.1 consists of one home country (or more than one, if legislation permits), but several countries of origin, several hosts and many markets. Within each country there may then be several interested parties to negotiate with on a wide range of issues. Furthermore, each bilateral negotiation may be dependent on, and in turn affect, the outcome of other negotiations.

Bilateral and multirelationships

It would be gratifying to imagine that if only the MNE could assemble all those with vested interests in one big conference to thrash out all the outstanding issues and agree on a common framework for action, the process of reaching agreements would be greatly simplified and beneficial to all concerned. But such a conference would be impractical for several reasons. First, many of those involved prefer bilateral negotiations, where each negotiator tries to pursue his own interests without having to compromise or be inhibited by the needs of third parties. Second, the negotiations are often kept confidential and the MNE may find that if the outcomes were to become public knowledge, it would inevitably lead to further concessions to other parties. Third, changing circumstances constantly require adjustments in some agreements, and if a conference were to be called in every case, the administrative machinery would become too unwieldy and each conference would be exploited by some of the parties as an excuse for renegotiations. Another important stumbling block is that each country has to contend and negotiate with many MNEs and for reasons of expediency alone (leaving aside the arguments cited above) it would not be willing to participate in a conference called to consider the problems of every individual MNE.

In addition to the relationships between the MNE and the four country-types, there are six other bilateral country relationships shown in Figure 10.1, concerning a wide range of economic and political issues. These cover all aspects of trade, including tariffs, import and export restrictions for goods and money, employment of foreign nationals, balance of payments, off-set arrangements, deals on exchange rates, treaties regarding relief on double taxation, agreements on patents and copyrights, and so on.

Interests common to many countries have led to many international conferences, without the overt participation of MNEs, to discuss global trade issues. One prominent example, which is referred to in Chapter 8, is the conference under the auspices of GATT (General Agreement on Tariffs and Trade), which has been active since 1948 in its efforts to reduce trade restrictions and which has been convened periodically to review and update international agreements. The eighth round, which began in 1986 and was due to be completed in 1990, was more ambitious than its predecessors, going beyond visible trade to cover trade-related aspects of intellectual property, as well as trade and financial services, while aiming at the same time to ensure the effective regulation and integrity of the financial markets. The main issues in the current round are textiles, agricultural products and property rights. Progress in the field of textiles would benefit the LDCs, progress on agricultural products is expected mainly to benefit the US, while progress on intellectual property rights should benefit the US and Europe.

Another example is UNCTAD (UN Conference on Trade and Development), where several developing countries have combined to form a pressure group vis-à-vis the industrialized countries, demanding unilateral tariff reductions to allow easier access to markets in the developed countries, availability of easier credit facilities and amelioration in the servicing of debts. All such conferences and discussions are conducted at a governmental level, though their outcome affects the scope of operations of many MNEs and the relationships shown in Figure 10.1.

Consequently, the process of the many bilateral negotiations in this model remains complex and messy. It often gives rise to misunderstanding and to entrenched positions, with each party determined to get the best deal at the expense of everybody else. Reference was made in Chapter 9 to the potential conflict of interests between the home country and the host country. As Figure 10.1 suggests, this conflict extends to all the countries involved, and the MNE is often caught in the middle, trying to satisfy many masters simultaneously.

As briefly indicated in previous chapters, the MNE is often accused of misusing its enormous economic power and of taking unfair advantage of the weakness of origin and host countries. Some MNEs have even been charged with political interference, for example ITT allegedly conspired with the CIA to stop the election of Salvador Allende in Chile in the early 1970s, and similar charges have been levelled against the activities of other American MNEs in other countries in Latin America. The criticisms voiced against MNEs are clear symptoms of the underlying conflict of interests, provoking a series of measures and countermeasures that exacerbate the situation further, as we shall see below.

Goals of the five players

The basic reason for the players taking part in the international trade game, and for being prepared to tolerate the messy relationships implied in Figure 10.1, is that each player stands to gain from the transactions involved. Their overall respective goals and some of the more specific means of attaining them are summarized in Table 10.1.

Note that some of the goals in the table are formulated as positive, in that they describe attainments generally regarded as highly desirable, while others are negative, representing risks and outcomes that should be avoided. Clearly, with any multigoal system, the various desiderata are rarely of equal weight, and some may well be incompatible with each other. There has been no attempt in the table to produce a comprehensive list of goals, to rank them, or even to suggest that they apply equally to all players of a given type (not all MNEs have identical goals, nor all home countries, all host countries, etc.). The purpose of the table is merely to outline the general direction in which the players wish to progress, and thereby indicate the potential areas of common interests between MNEs and the countries concerned, as well as the issues where interests are likely to diverge and generate conflict.

Clearly, the governments of the four country players share many general aims. They all want to achieve a positive balance of payments, high levels of employment, high living standards for their citizens, and high tax revenues in order to pursue their proclaimed social and economic policies. They all want to develop and strengthen their industrial base and to enjoy the fruits of modern technology. They all want to protect the environment and avoid pollution (though some countries merely pay lip service to these issues and seem less concerned about them than others). They are all fearful of loss of jobs to another host country, and of loss of tax revenues as a result of financial manipulations by MNEs.

In addition, the individual country players have aspirations of their own. The home country is anxious to ensure the repatriation of capital, profits, dividends and licence fees. It wishes to maintain a technological advantage over other countries and to avoid the loss of highly-skilled personnel (who would otherwise be usefully employed in industry and commerce in the home country). Origin and host countries often have identical goals. They seek to attract foreign investment, to establish an expanding industrial base, to gain foreign currency from exporting materials or manufactured goods, and at the same time they are determined to preserve their political and economic independence (many examples of attracting foreign investors are mentioned in earlier chapters). The market country shares some of these goals and is concerned to ensure that imports do not harm local industry (and thereby cause job losses), that

Table 10.1 Goals of the five players in the trade game

The MNE	Home country	Origin country	Host country	Market country
Profitability, i.e. attain desirable performance criteria (profit, revenue, ROI, profit margin, etc.)	Balance of payments	Balance of payments	Balance of payments	Balance of payments
	High living standard	High living standard	High living standard	High living standard
	Tax revenue	Tax revenue	Tax revenue	Tax revenue
	Employment	Employment	Employment	Employment
	Capital repatriation	High price for exports	High price for exports	Low price for goods
	Dividends etc.	Industrial base	Technology	
Size	Expansion of trade	Foreign exchange	Foreign exchange	
Product dominance			Develop exports	
Market dominance			Import substitution	
Low unit cost			Develop capital markets	
Low financing cost				
Independence				
Flexibility				
Avoid risks:	*Avoid:*	*Avoid*	*Avoid*	*Avoid*
Financial	Excessive capital outflow	Environmental damage	Environmental damage	Trade imbalance
Political	Loss of jobs	Pollution	Pollution	Damage to local industry
Regulations	Loss of tax revenue	Loss of tax revenue	Loss of tax revenue	Loss of tax revenue
Nationalization	Loss of technological advantage	Foreign domination	Foreign domination	Pollution
Foreign exchange losses	Loss of skilled manpower		Distortion of labour market	Monopoly
Economic uncertainty				Loss of jobs
Loss of control				Dumping of foreign goods
Cheap imitations				

customers get good value for money and have a choice of goods and services, and that an unacceptable drain on foreign exchange reserves is avoided.

The national interest and the MNE response

It is clear from Table 10.1 that government agencies in all countries are under pressure to act vigorously in their national interests. Of the many instruments of industrial policy that can be devised by national governments to that end, Bardach (1984) highlights the following ten for special consideration, and these have become prominent particularly in host and market countries:

1 Protectionism in the form of tariffs and other barriers.
2 Subsidies designed to promote exports.
3 Financial assistance to workers made redundant during the downturn of the product cycle.
4 Grants and subsidies to cover expenditure for innovation.
5 Investment in training programmes to improve the training skills of the workforce.
6 Unit-trust legislation and promotion of competition
7 Creating a capital infrastructure to help industrial development.
8 Providing loans and loan guarantees to industry.
9 Adopting a tax policy designed to encourage capital investment.
10 Implementing a government procurement programme aimed at supporting designated industrial sectors.

All these instruments are designed to strengthen the local industrial base and they inevitably affect the relationships between governments and MNEs.

As the economic and political climate changes, governments have to reassess past agreements with MNEs and try to negotiate better terms. In many cases governments can use their powers to impose new conditions, either to supplement the original deals or to modify them. The temptation to change the rules stems from the fact that once the MNE has established its operations in a particular country, its locked investments reduce its room for manoeuvre and it is then in a weaker position to resist governmental interference. In contrast, the host country becomes richer and wiser with the passage of time (particularly if it has had the benefit of negotiating successfully with other MNEs) and is in a much stronger position to bargain.

When the impact of governmental actions is relatively small, the MNE has to take them in its stride. There comes a point, however, when the

MNE has to retaliate, at least as a warning to the government (and perhaps governments of other countries) to refrain from taking actions in the future, which are likely to harm the MNE's interests.

Table 10.2 *Examples of governmental actions and MNEs responses*
Background: MNE transports materials and components from country A to country B, where manufacturing takes place, and exports the products to country C

	Government action	MNE's response
1	A increases price of materials	MNE seeks alternative source
2	A insists on joint venture	MNE seeks alternative source, increases transfer price to reduce profit
3	A resorts to nationalization or expropriation	MNE demands compensation, reduces investments, withdraws support and advice, seeks new source, organizes international reprisals
4	A or B increases tax on profit	MNE increases transfer price to reduce tax liability
5	A or B puts limits on profit repatriation	MNE manipulates transfer price to reduce declared profit in A or B
6	B imposes tariff on imports of components	MNE transfers manufacture of components from A to B
7	B imposes tariffs on imports of raw materials	MNE seeks another host country
8	B insists on technology transfer	MNE transfers low technology, establishes local R&D lab
9	B constrains capital outflow or the convertability of the currency	MNE uses profits to expand operations and to diversify investments in B
10	C imposes heavy tariffs on all imports of manufactured goods	MNE puts a manufacturing plant in C
11	C imposes heavy tariffs on imports from B	MNE supplies C from another host country
12	C imposes price controls on goods distributed by the MNE	MNE cuts costs, increases transfer price, reduces quality and product range, reduces RTD&D budget
13	C encourages cut-throat competition in order to reduce the price of goods	MNE increases efficiency, cuts prices, transfers production from B to C
14	B's currency is revalued compared with that of C, making imports too expensive for C's market	MNE transfers production capacity from B to C and raises capital in C to finance operations there

Some examples of the MNE common responses are shown in Table 10.2 for the case where A is the origin country, B is the host country and C is the market country. Some of the governmental actions depicted in the table are drastic, for example that of nationalization and expropriation of the MNE's assets, or substantial increases in tariffs and taxes that may jeopardize the future of the venture, in which case a robust response from the MNE is called for (some aspects of the imposition of tax policies were discussed in Chapters 5 and 8). In other cases, the profitability of the venture may be adversely affected, not to the extent that would justify it being abandoned altogether, but the damage may be sufficient to dampen any enthusiasm for making further investment in the country in question or to contemplate new ventures in that country. Furthermore, other MNEs may be discouraged from investing in that country if they judge the conditions to be less favourable as a result of governmental action.

In addition to the examples in Table 10.2, there are many other ways in which governments can obstruct the operations of MNEs, such as selective constraints on specific products, discrimination in government procurement, covert subsidies to local competitors, more exacting health and safety regulations, obstructive documentation and red tape, imposition of technical specifications and standards (including packaging and labelling), delaying the award of copyright protection and thereby exposing the MNE designs to ruthless imitators. All these are measures commonly found in countries wishing to frustrate importers, but they can also be employed against MNEs with plants and distribution systems in host and market countries, either as a means of support for local producers, or as a leverage to be used in future negotiations with the MNEs.

Another popular ploy by a market country (used, for example, by several countries with respect to imports of automobiles and aerospace products) is that of specifying a minimum of 'local content', to force the MNE to buy components from local manufacturers. The purpose of this provision is not only to support the local industrial base and safeguard a minimum level of employment, but to improve the balance of payments and foreign currency reserves.

At times, governmental action is sufficiently extreme to cause the MNE to withdraw altogether. One example is that of India, which in 1977 imposed a restriction of 40 per cent as the upper limit for ownership of local enterprises by foreigners. Coca-Cola and IBM were already well established in India and it became clear that the new decree would profoundly affect their operations. On Coca-Cola, India's Minister of Industry said:

> The activities of the Coca-Cola Company in India during the last twenty years furnish a classic example of how a multinational corporation operating in a low-priority, high-profit area in a developing country attains runaway growth and, in the absence of alertness on the part of the government

concerned, can trifle with the weaker indigenous industry in the process. The manufacture of beverages should be 'Indianized' and the outflow of foreign exchange should be halted (*New York Times*, 1978).

Similarly, IBM was accused of selling obsolete computers in India and generally importing low technology. Both Coca-Cola and IBM concluded that they could not relinquish control of their operations and both decided to quit.

Moves to obstruct foreign investors from owning local operations are not confined to underdeveloped or developing countries. Public utilities, such as water, gas, electricity and transportation systems, are sensitive services and their prices can have a profound impact on the cost of living. Consequently, control of such utilities by foreign companies is prohibited in many countries. Similarly, mineral resources in some countries are state-owned and in others ownership is strictly confined to indigenous corporations. In several countries the control of the media (newspapers, television and advertising) by foreign nationals is severely constrained.

Even in highly industrialized countries, ownership of local enterprises is not necessarily freely open to foreigners. In Canada, the sensitive issue of foreign domination exercises many politicians and the provincial governments have wide powers to reduce foreign ownership of mineral resources in the mining industry. In Japan, rules restrict the extent of shareholdings in many industries. In Britain, the ruse of the 'golden share' has been used by the government 'in the national interest' to prevent denationalized companies from falling into the hands of foreign predators. For example, foreign ownership of shares in the Jaguar motor company had been strictly limited by government decree for many years, before the government eventually relented and allowed Ford to take it over in 1989. Another example is the case of Rolls-Royce aeroengines, where foreign share holding is limited to 29.9 per cent.

Even in the US, regarded as the bastion of free market enterprise, there have been strong moves from time to time to keep foreign enterprises at bay. The influx of foreign investors into the US, particularly from Europe and Japan at times of a weak dollar in the foreign exchange markets, has alarmed many American politicians. Lord Hanson of Hanson Trust pointed out in 1990 that thirty-nine states in the US had some form of legislation to prevent takeovers. In some industries, such as banking, forays by outsiders to gain control of local banks have been frustrated by a web of intricate constraints, some designed to prevent monopolies and 'to protect the public'.

A more recent, and perhaps more telling example, has been the $1.6m bid by the British industrial conglomerate BTR for Norton, the US abrasives and ceramics company based in Massachusetts. The takeover battle provided a field day for local and national politicians. Senator Edward

Kennedy of Massachusetts strongly opposed the bid and said: 'We need more Nortons, not subsidiaries of distant conglomerates who can only distinguish between a grind wheel and a tennis racket by looking at a balance sheet.' Congressman Joseph Early from the same state was more specific: 'We are particularly alarmed over the possible fate of the research and development [which Norton] have conducted over the last twenty years – research and development we believe is vital to our national security.' Some 120 members of the Congress urged President Bush to set up a security investigation of the bid (under the 1988 Trade Act the President is obliged to review foreign acquisitions of US companies involving security considerations) and other politicians also urged the government to intervene 'on behalf of a great American company and its American workers'. The Governor of Massachusetts, Michael Dukakis, duly obliged early in 1990 by rushing through the state legislature a new law restricting to one-third the number of directors of a company that can be removed from the board in one year. Referring to the US struggle for independence over two centuries earlier, he then proclaimed triumphantly: 'Another war of independence, another attempt by a foreign power to interfere with or attempt to shape, our destiny – 215 years later – and we won again' (*Observer*, 1990). It is ironic that the strong opposition of Norton to be taken over by a foreign predator in the shape of BTR should have culminated shortly afterwards by the Board agreeing to the takeover by Compagnie de Saint-Gobain, the French industrial conglomerate.

The effect of trade unions

The actions shown in the examples in Table 10.2 are all government induced, all compelling the MNE to take stock and consider its position. At times, however, the MNE may feel obliged to respond to other threats and actions, such as those from trade unions. In 1988, for example, Ford decided to abandon its plan to build a £40m electronics plant in Dundee in Scotland, after failing to achieve a single-union agreement with the trade unions. Ford felt that its experience with multiunion plants underlined the threat of incessant union rivalry and labour unrest. In 1990 Ford cancelled another big investment plan in Britain, this time for a £225m engine plant in south Wales, which would have been a key supplier of engines to other Ford plants in Europe. The decision to move this investment to Germany was taken in spite of the relative low level of wages in Britain (to some extent offset by the higher labour productivity in Germany). But in the wake of a series of unofficial strikes (which, it was claimed, cost the company £10m a day) the Ford management concluded that, 'particularly in the light of the unreliability of supply we have experienced in our British plants in recent years' (as the company spokesman put it), it could not take the risk of disruptions and their domino effect through the other European plants.

Although it deeply deplored the loss of potential jobs, the British Government was powerless to intervene in both cases, since all the negotiations were conducted between the Ford management and the unions. The Government expressed the hope that the trade unions would learn from this experience and realize that in the international trade game national frontiers have lost their significance when it comes to siting manufacturing facilities.

International corporate networks

Needless to say, the British trade unions were furious with the Ford decision and would no doubt seek in future to establish alliances with European trade unions in order to form a united front in their negotiations with MNEs such as Ford. In this case, however, Ford can take comfort from the fact that the German trade unions would be keen to ensure that the investment does go to Germany and their declared support for a united front would be more a symbol of solidarity in principle, rather than a commitment of substance. Playing one country against another is still one of the effective ways for the MNE to achieve its purpose in terms of cost, flexibility of operations, retaining control, and minimizing disruption of supplies.

The power of the MNEs has naturally provoked concerted action by those affected, as demonstrated in the area of price fixing and restrictive output allocations exercised by mineral producing countries. OPEC (the Organization of Petroleum Exporting Countries) is perhaps a prime example of a producer cartel attempting to dictate the price of crude oil and to allocate production quotas to its members, though compliance with OPEC's decisions remains voluntary. Producers of minerals and natural resources, such as copper, chromium and rubber, have tried to copy the example of OPEC, so far with limited success, though such efforts will undoubtedly continue.

Political issues

Much of the discussion concerning the conflict between MNEs and host countries (which in this context includes origin countries as well) has centred on economic issues. The trade game is depicted as a series of activities conducted under certain agreed rules, where all the players stand to gain more from participating than from abstaining. As Robock and Simmonds (1983) put it: 'International business activity is not necessarily a zero-sum game in which one nation has to lose in order for another to gain. If international business results in a more efficient use of world resources, all parties can secure increased benefits. Because the pie to be divided is larger, each nation can have a larger slice.' But the fact that the trade game produces wealth does not deter the parties from squabbling about how it should be divided. Naturally, each player tries to get the best deal, even when an improved reward for one clearly means a diminished reward for another, and the final distribution reflects the relative power of the parties and their incentive to continue to participate.

All these economic issues need also to be reviewed against the major political considerations of the parties concerned. One of the main worries of an MNE in this respect is whether the host country, particularly if it is an underdeveloped or one of the emergent developing countries, has the necessary political stability to safeguard the future of an investment. The risk of nationalization, expropriation, one-off punitive levies, arbitrary changes in relevant laws, unilateral cancellation of agreements, and plain disruptions owing to political unrest, have all contributed to the MNE's reluctance to invest heavily in such countries. Potential investors have plenty of examples of political instability resulting in the overthrow of national governments by freedom movements and the emergence of new regimes hostile to foreign investors for reasons of nationalism and ideology. Drastic changes of policy towards foreign Western investors are not uncommon, as suggested by the examples of Egypt, Lybia, Syria, Cuba, Chile and others. In some countries, kidnapping and blackmail have become rife (whether for terrorist or criminal motives), such as in Argentina, Columbia, Venezuela, Mexico and Lebanon, and even in Europe, as in the case of Ireland and Italy. The fear for the safety of their executives must understandably inhibit many MNEs from pursuing investment opportunities in countries that exhibit characteristics of political and social instability.

The increasing tendency for risk averseness on the part of MNEs, even with respect to stable countries, has prompted host countries to offer attractive incentives to potential investors (as indicated in Chapter 9), such as tax holidays, special development grants, protected markets for a specified period, subsidies for training and infrastructure development, low interest loans and so on. Further to these financial incentives, host countries are now willing to enter into 'foreign investment guarantee agreements', which spell out detailed procedures (including a mechanism for independent arbitration) for settling all disputes, with firm promises of fair and prompt compensation against losses due to wars, currency devaluation, or expropriation.

In addition, MNEs have been apprehensive about the bias of local courts and the fact that international law has often proved to be weak and inadequate. To reassure MNEs on these issues, international bodies have been set up, prominent among them is ICSID (International Center for Settlement of Investment Disputes) established in 1980, its decisions being final and binding on all member states. Many states subscribe to ICSID, though some countries object on the ground that it is an infringement of their sovereignty (Robock and Simmonds, 1983). As more countries come into the fold, and as MNEs show increased determination to deal only with ICSID members, some of the mutual suspicion and mistrust is likely to diminish. Other international agencies have concentrated on agreements to regulate areas which have a significant effect on business and trade, such as air and sea transport, telecommunications and postal services, health and safety, trade union rights, pollution control and preservation of the environment.

Nonetheless, the paramount concern of host countries to safeguard their sovereignty remains. This found expression in the UN Charter of economic rights and duties of states, passed by an overwhelming majority (with very few dissensions and abstentions) at the UN General Assembly on December 12, 1974: 'Every state has and shall freely exercise full permanent sovereignty, including possession, use and disposal, over all its wealth, natural resources and economic activities.' This declaration, strongly supported throughout the developing world, is interpreted by many as giving host countries absolute authority over their economies and natural resources, with their national laws having the final judgements on questions of nationalization, compensation and termination of agreements (Robock and Simmonds, 1983). In South America this approach has been reinforced by the so-called Calvo doctrine (named after an Argentinian lawyer), stating that a foreigner entering a country is automatically subject to its laws and cannot seek the protection of exterritorial agreements. No wonder MNEs get jittery.

Transfer pricing

It is natural for the management of the MNE to strive to maximize its profits and minimize its tax liability throughout its operations in various countries. For example, if ad valorem duty is imposed on imports of raw materials and components, the MNE would endeavour to assign a low value to imported inputs. Similarly, if corporation tax is linked to the level of declared profit, then increasing the cost burden and minimizing profit automatically reduces the local tax liability.

Frequent references are made in Table 10.2 to transfer pricing as one of the ways in which the MNE can manipulate its costs and declared profit in any given country. If the MNE operates in several countries, it is in its interest to declare the least profit in a country of high corporation tax and to shift the profit to a country where the tax is lowest. This is achieved by the transfer price charged by a subsidiary in one country for goods and services supplied to a subsidiary in another country.

It is not difficult for an MNE to set up a sequence of international transactions through a network of subsidiaries, some of which may in practice be no more than invoicing and documentation processing units, specifically established in low-tax countries for the purpose of siphoning profits from high-tax countries. A further advantage of diverting profit from a host (or origin) country is that it gets around the problems of exchange controls (where such controls exist), leaving aside the probability that a low profit-making venture may make it less attractive to expropriate.

But, as pointed out in Chapter 8, governments are getting wiser in their understanding of international trade operations and managerial manipulations. As economic conditions and factor input prices constantly change in the market (including the prices of basic commodities, the cost of labour and the cost of capital, and hence the cost of components and other inputs), the prevailing prices at the time of signing the agreement can hardly form the basis for future determination of transfer prices. Negotiations between an MNE and a host country often involve elaborate stipulations regarding accounting conventions and the definition of transfer prices, the procedures to be followed in future determination of these prices and the degree of freedom allowed to the MNE to deviate from laid down specifications. The agreement, therefore, needs to specify the formulae that would produce the ingredients of future computations.

Two methods are often used. The first relies on price indices for the major inputs, which sounds simple in theory, but is quite difficult in practice. Apart from the question of whether to rely on national or international indices (and these can diverge significantly with the passage of time), there are always problems surrounding the construction of any particular index, its continued stability and relevance, and the degree to which it can reflect changes in local conditions that affect manufacturing and administrative costs. It is often argued, however, that no index is perfect, but that an agreed index is better than no index at all. Experience suggests that while national price indices (for example to measure inflation) are commonly used within a given country, they are less acceptable across national frontiers.

The second method is based on the proposition that the subsidiaries in the different countries should operate and interact on an arm's-length basis, as if they are not obliged to buy and sell from each other but are free to negotiate in the open market. The concept of an arm's-length relationship has the attraction that the price of goods flowing across the border can then be ascertained 'objectively' by the market price and is not subject to the whims and manipulations of the head office in the home country. In practice, however, it is difficult to determine the market price without having concrete quotations from competing suppliers, and these may neither be genuine nor easily available.

In addition to the transfer price of physical goods, there are many other charges that can legitimately be levied on a subsidiary company, such as the cost of RTD&D, licensing fees, promotion and marketing, training and skill transfer, interest charges on direct loans and other financing costs, and finally head office direct and indirect costs associated with controlling the subsidiary. All these charges are less amenable to assessment by the so-called 'market price', so that even when rigid formulae for the value of physical supplies can be agreed, there are many ways in which the MNE can still manipulate the overall transfer price in pursuit of its aims.

The role of joint ventures

Partly for political and partly for economic reasons, some host countries insist that an investment by an MNE takes the form of a joint venture with a specified minimum of local participation. In some cases the minimum is set at 51 per cent of the shareholding, in others the regulations indicate the degree to which managerial control over operations and profit plans must be held in local hands.

Many MNEs refuse to operate under such conditions and would not contemplate a venture in a host country, unless it is a fully-owned subsidiary. However, there are countries in which no such constraints exist, and yet the MNE prefers to participate in a joint venture for a variety of reasons:

- The locals have a better knowledge of the local market and how it can be developed.
- A joint venture reduces the financial risk to the participants, compared with that of the single investor.
- Local participation reduces the political risk of hostile government action, such as expropriation or punitive taxes, and ensures that the local partner would lobby the government and local politicians in pursuit of his interests.
- A joint venture is bound to rely partly on local management and therefore reduces the need to import valuable (and often expensive) manpower from the home country.

- A joint venture may open opportunities for raising local finance for working capital and possibly for expansion and diversification.

There are, however, many disadvantages. Leaving aside the need for the MNE to share the profits of the joint venture with a partner, there is the loss of complete managerial control of the operations. The existence of a partner means that the MNE is not free to make unilateral decisions about expansion, takeovers, disposals, reorganization or financial restructuring of the business, appointment of key personnel, or even changes of orientation of products and markets. All these strategic issues need the agreement (and, at times, financial participation) of the partner, and the required consultations may prove cumbersome and time-consuming.

In such circumstances, the joint venture can neither become an integral component of the MNE's overall structure, nor can it play a part in the MNE's long-term strategy in the same way that a wholly owned subsidiary can. In a sense, the joint venture then falls into the category of a foreign investment, except that the MNE may easily get locked-in, when it discovers that it is far more difficult to dispose of its share of the business than if it were fully owned. Thus, the main advantage of engaging in a joint venture remains that of reducing many of the risks implied in Table 10.2, but this advantage needs to be set against the loss of control and flexibility. A further problem is created when the joint venture starts developing its own RTD&D capability, and if that leads to new inventions and new designs, their potential benefits to the parent MNE may only be realized through arm's-length arrangements, whereas a fully owned subsidiary would allow instant access at no extra costs.

In spite of all these disadvantages, the joint venture is an effective means of gaining acceptance and respectability in the host country, in the eyes of both the government and the general public. The accusation that the MNE is an imperialist tool of exploitation loses credibility when the joint venture demonstrates responsible behaviour and concern for the welfare of the host country and its citizens. Instead of acting as a foreign investor, who is only interested in a quick buck, the MNE needs to proclaim its long-term intentions and commitment. True participation by local businessmen and local capital may be the best way to ensure that commonality of interests is widely acknowledged, so that many of the potential conflicts alluded to earlier would simply disappear.

Code of conduct

In a less congenial environment, as suspicion, mistrust and greed grow, the trade game can clearly degenerate into open warfare between the

MNE and national states, and in some cases has already done so. The need to create the conditions for all the players to behave honourably and for mutual trust to be established has long been recognized by all concerned. There is obviously a strong case for a code of conduct, both for the MNE and for the host country (Grosse, 1980).

The OECD has published its suggestions in this area (see for example OECD, 1976) and the UN has convened many commissions and conferences on these and allied themes. These moves, aimed at reaching an agreement on the control of international operations, have generally been widely welcomed by national governments and by MNEs, though some commentators have serious doubts about the effectiveness of what has been achieved so far and point to the 'different conceptions of who is going to be controlled, who is going to do the controlling, and what the purpose of the control will be' (Bergsten et al, 1978).

Host countries may argue that the development of machinery such as INSID (International Centre for Settling of Investment Disputes, mentioned earlier) and other international conventions, coupled with binding bilateral agreements, should give sufficient reassurance regarding the expected conduct of sovereign states and that MNEs should respond with a binding code of conduct that would command respect among host countries. Some MNEs have indeed tried to formulate proclamations to that end, but others fear that such a gesture would merely make them hostages to fortune and weaken their bargaining power in future bilateral negotiations. They argue that the long history of unilateral drastic actions taken by states gives MNEs little confidence that international compacts will hold, and that in the light of widespread political instability the MNEs should continue to be wary and primarily look after the interests of their shareholders.

Nevertheless, there is so much to be gained from the cessation of hostilities and from creating a more cooperative investment environment, that every effort should be undertaken under the auspices of the UN, the World Bank and a newly-created association of prominent MNEs to formulate codes of conduct. If such codes were to be agreed, and if both countries and MNEs would increasingly refrain from dealing with parties not signatory to these codes, then pressure would build up for the codes to be widely accepted as the rules of the international trade game, from which all participants would continue to profit. The establishment of such codes, coupled with a more aggressive policy to pursue joint ventures, would transform the scope of international operations.

References

Bardach, E. (1984), 'Implementing Industrial Policy' in Chalmers, J. (ed), *The Industrial Policy Debate*, Institute for Contemporary Studies, San Francisco.

Bergsten, C. F., Horst, T. and Moran, T. H. (1978), *American Multinationals and American Interests*, Washington Brookings Institution.

Grosse, R. E. (1980), *Foreign Investment Codes and the Location of Direct Investment*, Praeger, New York.

New York Times (1978), 2 June.

Observer (1990), 22 April.

OECD (1976), *International Investment and Multinational Operations*, OECD, Paris.

Robock, S. H. and Simmonds, K. (1983), *International Business and Multinational Enterprises*, Irwin, Homewood, Illinois.

11 Concluding remarks

> - The race
> - Innovation and 'critical mass'
> - 'Betting the company'
> - The rise of the MNEs
> - The question of government support
> - A realistic policy
> - Support for university research
> - MNEs and sovereign states
> - A code of conduct
> - The global issues

The race

The rapid changes in technology in recent years are at least as significant as the first industrial revolution and pose a serious challenge both to industry and to national governments. There is a relentless race to introduce new products through the exploitation of technology as a source of economic advantage. It results not only in opening entirely new markets, but also changing the whole nature and structure of many industrial sectors. The areas of computers, information technology and telecommunications, aerospace and pharmaceuticals are striking examples of the effect of advances in modern technology, sweeping aside existing and well-entrenched product ranges. This is why innovation has become a central ingredient and a moving force of corporate strategy. No leading enterprise can afford to be left behind, since the technology race may well spell extinction for the laggards.

To remain in this race, the enterprise must not only be in command of the technologies involved, but have the managerial skills to convert them into commercially successful products. This is accomplished through the integration of design, production and marketing more effectively than the competition. Appreciation of this theme, which we sought to explain in Chapter 1, is absolutely essential to the understanding of the consequences both for industry and governments. Invention and the demonstration of feasibility, vital though it is as the creative input to innovation, must not be mistaken for the whole of the process. More is required for success. In terms of the cost of launching a new product it may amount to as little as 10 per cent of the total costs. But to realize commercial success the management skills of executing and integrating design, production and marketing are equally important. Shortcomings in these areas may well result in the failure of a potentially very innovative idea.

We emphasize this point because of continuing misunderstanding in many quarters of this important feature of the innovation process. This misunderstanding manifests itself when 'R&D' is loosely equated with innovation and consequently when R&D expenditure is taken as the measure of the innovative activity of an enterprise. It also manifests itself when government subsidized and managed innovation fails commercially due to lack of integration with production and marketing skills, coupled with the absence of investment in these areas.

Innovation and 'critical mass'

The key element of an innovation strategy lies in the product cycle, which is discussed at some length in earlier chapters. At the beginning of the innovation and design process there is a negative cash flow, which can continue for several years and its cumulative effect can be very substantial where high technology is involved. The considerable costs of launching a new product, which amount to a large sum compared with the cost of the underlying invention, can only be recovered through sales, and this fixes a minimum size for the enterprise which is able to undertake such a launch effectively. Another important influence on the size of the enterprise is that for any number of reasons, not by any means all under the control of management, a new product may fail. The probability of failure is quite significant, and possibly only one in six projects may prove to be successful, and this means that the enterprise must be able to accept this level of expenditure without the risk of going out of business.

To evaluate the order of magnitude involved, a simple formula for the costs is proposed in Chapter 2. Typically, it gives figures of about £2.5bn for a popular motor car, £2bn for a new generation mainframe computer, about £750m for a new camcorder, and some £3bn for a medium-sized airliner (150 seater) to replace those currently in service. The stakes are, therefore, very high and many have likened the macroinnovation process to a gamble, where 'betting the company' means that the winner takes a substantial slice of the market and uses the proceeds to prepare for the next round of battle. Enterprises risking such large sums of money must be commensurately large and have sales outlets capable of recovering them.

Both the size of the investment and the time involved for recovery, which can be fifteen years or longer, warrant a new name for such industries, and we have adopted the prefix 'macro' to identify both the industry and the innovation involved. *Macroinnovation* means risking the order of a billion pounds or more in a new product. The arena in which this kind of competition is conducted is the *macroindustrial economy*.

The emergence of the three major trading blocs – the USA, Japan and the EC – and the increasing preoccupation with global markets have further underlined the pressures of competition and the trends described earlier. This pressure manifests itself in the move to establish enterprises of a scale capable of competing in the global race and their development is proceeding at an ever-accelerating pace. As shown in an earlier chapter, in 1989 there were some seventy-five companies, roughly divided equally between the three blocs, all with sales of £4bn or more per annum. In the motor industry not only are the sales of the prime manufacturers much larger than this, but their suppliers are now also actively regrouping themselves into macroindustrial units.

Given that an enterprise needs to have a 'critical mass' in order to compete in a given industrial sector, the enterprise can proceed along several possible strategies. One is to be first in the innovation race to become the lead macroinnovator, well ahead of the competition; another is to lag somewhat behind and let the trailblazer make all the initial mistakes and incur the corresponding very high costs, but plan to move in rapidly and competitively when the pioneer shows signs of being successful. The first is a much more costly and risky strategy, but the winner can reap very rich rewards. The second reduces the expected immediate financial benefits but also the enormous risk of failure. Industrial history records many examples of the success and failure of both strategies.

The interesting case of video recorders is a good example of the second strategy, where Sony launched its Beta-max product while the rewards of this macroinnovation went to those who came second with the VHS system. Currently, civil aviation has two inventions presenting this strategic problem – the prop-fan engine for future long-range air travel and the tilt-wing vertical take-off aircraft for short haul routes, obviating the need for large airfields which add excessively to journey times (e.g. between city centres). The risk that initial problems would be damaging and costly in both cases needs to be set against the possibility of immense rewards if the market takes off. Since the price of fuel is an important factor in the first case and political attitudes to transport policy are crucial to the second, it is clear that the risks are by no means confined to the technical competence of the companies involved.

'Betting the company'

The financial resources needed to play the 'betting the company' game are clearly enormous. There are three ways in which an enterprise can alleviate the level of risk involved:

1. The first is to be sufficiently big to be able to diversify the corporate innovation and product portfolio, so that by being involved with several product cycles at different stages of their development, the overall cash flow (and the overall risk) can be more effectively controlled.
2. The second is to seek financial assistance from a national government, arguing that the country cannot afford to fall behind in the technological race, which affects not only the enterprise in question, but also suppliers and allied industries, with serious implications for employment and the balance of payments. Many national governments have been persuaded by these arguments and have poured vast sums of money in the form of direct or indirect aid to subsidize the innovation effort. More recently, the CEC (Commission of the European Communities) has embarked on an ambitious programme to support innovation on a massive scale in several areas, information technology and telecommunications being the most notable (as discussed in Chapter 7).
3. The third is to share the risk with other enterprises through alliances and joint ventures, and this route has become increasingly popular in recent years, as the many examples cited in earlier chapters suggest. The reason for this trend is related to the fact that stakes involved in the macroinnovation process have been rising steadily, coupled with the reluctance of some governments to commit themselves to innovation expenditure on a massive scale, and also to the desire of enterprises to become multinational and independent of national governmental intervention.

Joint ventures have a superficial attraction to governments as they appear to preserve national champions in what is really a supranational industrial scene. They may also appear to be more attractive to governments than the alternative of having to provide direct or indirect assistance to innovating companies. It is in this area that many challenges to governments arise, torn as they are between the free market principle of non-intervention and the manifest importance to the national wellbeing of having a significant share in the innovative industries of the world.

The rise of the MNEs

All this explains the need for enterprises, aspiring to play a role at the commanding heights of industry and business, to be big and to become multinational. It is only through size, as indicated in Chapters 3 and 4, that enterprises can hope to have sufficient financial resources, skilled manpower, as well as technological and managerial capabilities, to become players at the macro level. The other reason for enterprises becoming big is the effect of economies of scale in manufacturing and distribution, and these advantages are discussed at some length in Chapter 4.

'Small is beautiful' is a well-argued thesis in the literature, but for the reasons discussed in this volume, big is necessary and consequently the organizational challenge in managing large organizations is considerable. Which functions to centralize and which should be set up as autonomous units has become a key organizational issue, in which modern information technology will play an increasingly important role. This issue is already so important to some macroindustrial enterprises that they have abandoned the vagaries of national telecommunications systems for dedicated satellite-based systems to cater for their own requirements, and this has resulted in significant implications for the way the business is run and organized.

All these developments, and the rise of MNEs (multinational enterprises), have focused attention on their role and on their relationships with national governments. The international trade game, which requires raw materials to be procured in origin countries, transported to host countries as inputs to manufacturing plants, and their products distributed in market countries, involves a complex range of operations, which the expertise and resources of the MNE can undertake.

While this positive role is widely acknowledged, national governments are suspicious of MNEs' motives for a variety of political and economic reasons; chief amongst them is the fear of loss of control and its implications for sovereignty. This concern is not confined to the many countries which are outside the three major trading blocs, but also within these blocs themselves. Every country wishes to have a thriving economy, and the success of its industry can be both enhanced and impeded by the operations of MNEs. No wonder the seeds of conflict between MNEs and national governments have resulted in many confrontations, in some cases exacerbated by nationalization and expropriation of MNEs' assets. Political instability in many countries has added to the reluctance of MNEs to commit themselves to further investment in certain parts of the world, thereby only fuelling the hostility of the countries involved.

Two fundamental questions emerge from the discussions in this book. The first is the challenge of macroinnovation, in terms of sponsorship and financing. The second is how to resolve the inherent conflict in the relationships of MNEs with host countries. The two questions are interconnected, since the establishment of global markets is a necessary condition for macroinnovation to take place, while at the same time the interests of national governments, both in home and host countries, cannot be pursued without these governments maintaining control of their national affairs. These interests, however, cannot be divorced from developments taking place in MNEs and therefore from the outcomes of the innovation process.

The question of government support

The above discussion puts into sharp focus the question of whether governments should consider providing direct financial support to MNEs in the form of overt assistance for undertaking innovation in designated areas, or indirect support through development grants, low-interest loans, tax concessions and other covert benefits.

For many years companies operating in high technology industries have looked to governments for support. Prime examples can be found in the USA, Britain, France, Germany and Japan in such areas as computers, telecommunications, aerospace, transport, shipbuilding, machine tools and other engineering sectors. In some industries the support was imbedded in a deliberate policy of public ownership, in others the policy was confined to substantial government grants, with varying degrees of control and intervention. Proponents of the market economy have been strongly critical of this policy, arguing that the performance of public sector enterprises has been poor compared with that of private industry, and under the Conservative Government in Britain many public enterprises have been denationalized or privatized during the 1980s.

The failure of many government-inspired large-scale projects to come up to expectation in economic terms, such as the Concorde and other ventures in the aerospace industry, have been heavily criticized. Many politicians and commentators in Britain have long concluded that the government is simply incapable of managing such projects in accordance with normal commercial criteria. Similarly, they have argued that financial support for innovation should be severely curtailed, or abolished altogether, leaving industry to make a judgement as to what scope and level of innovation it should undertake.

Consequently, proponents of the free market have urged all governments to conspicuously abstain from controlling or participating in large-scale projects and to withdraw overt support for industrial innovation. Some governments, including the British Government, have been sympathetic to this policy, though direct and indirect government involvement in certain countries, such as Japan, Germany and the USA, continues to play an important role, at least in providing a framework within which industrial development and macroinnovation can take place.

Nevertheless, free market economists and politicians are adamant in espousing a strictly non-interventionist policy. They strongly argue that industry should learn to stand on its own feet and finance all its needs, including those of innovation, from its own resources, or from resources raised in the capital markets. We have already drawn attention to escalating financial outlays required for macroinnovation, emanating from advancing technology and the pressure of global competition. The prospects of enterprises being able to rely on internal financing for this purpose are becoming increasingly remote, and even for the very few that can afford the expense, the risks are uncomfortably high.

Reliance on raising finance in the capital markets is then inevitable. The trouble is that the financial markets in both Britain and the USA are risk-averse and have become addicted to the philosophy of short-termism. Any proposition that does not meet demanding financial criteria for return on investment, which emphasizes the importance of the bottom line in the short term, is likely to be rejected by financial analysts and investors. This attitude is contrasted with that encountered in Japan and Germany, where long-term prospects are of great importance, so that modernization and innovation are regarded as long-term investments and are looked upon much more favourably than in the USA and Britain.

Admittedly, one way of financing macroinnovation is through industrial alliances and joint ventures, and we gave many examples in earlier chapters of this trend. This solution is attractive to many corporations, but it is not without its danger. The first is the loss of identity of the constituent enterprises and the creation of megaorganizations outside the control of democratically-elected national governments, and which are perceived to be answerable to nobody. As an extreme scenario, such alliances could lead to the rise of monopolies, curtailment of customer choice and arbitrary disregard of employees' needs in selected countries.

Clearly, the many vexed issues facing industry and the financial markets go beyond the unresolved debate among commentators about short-termism and are at the heart of the issue of government intervention. When the House of Lords Select Committee on Overseas Trade recommended in 1985, having taken evidence from a wide range of industrialists, that the Government should take certain actions to arrest the unquestionable decline in the British manufacturing sector, the report was greeted with a vicious attack from the guardians of the free market and the sacred memory of Adam Smith. The heading of the article in the *Financial Times* 'Coronets and Begging Bowls' captures the tone and level of the response to the report, dismissing the notion of Government intervention as sheer heresy.

In contrast, the response of the Fellowship of Engineering in Britain to the problems facing industry in the new technological age may be encapsulated by the prestigious and thought-provoking symposium organized in 1985 on the subject of 'Man and technology'. Every aspect, good and bad, of technology's impact on life on the planet was examined. In this context, it is interesting to quote Robert Frosch, then Vice-President (Research) of General Motors, who gave a paper in which he regretted the failure of non-engineers to appreciate the way technology is changing society, and he went on to say (Frosch, 1983):

> I find it interesting that the meetings which discuss this subject, like the present meeting, are almost entirely stimulated, called and populated by engineers and scientists. I have been trying for some time, along with other colleagues, to get lawyers, the economists and the political scientists at least marginally interested in this long-term question, but they seem entirely occupied in facing the past.

A realistic policy

We believe that a strict non-interventionist policy by government is quite untenable. We further believe it is in the interest of national governments to have a form of participation (through a consensus embodied in a politically independent joint government-industry agency) in innovation financing, without intervention in the management of the process, for four main reasons:

- To encourage a longer-term perspective than the narrow demand for bottom-line performance in the short term.
- To ensure that the fruits of innovation are available to strengthen the industrial base and the competitiveness of industry as a whole.
- To arrest the process of alienation of large companies from their home countries, a process that is bound to accelerate as internationalism and rootless corporations continue to develop.
- To balance the support, direct and indirect, granted to competitors by other governments.

From the perspective of the enterprise, such participation would have the assurance of support and continuity in the development of the innovation process, as suggested by the policies adopted in Japan.

Ensuring that such participation does not become a mindless subsidy for sleepy enterprises is a prime challenge. Generally, the mechanisms for the purpose do not exist in Britain and the USA. A special cadre of high-calibre officials with high mobility between the civil service and industry, on the lines adopted in France, would be a necessary condition before the appropriate control mechanism could be developed.

Support for university research

Also, some of the government support for research would need to be channelled to encourage long-term research in the universities and in research institutions. The fact that many British inventions have not led to commercial exploitation has convinced politicians that British universities are completely lacking in commercial acumen and do not understand the latter parts of the innovative chain described in the early part of this volume. This observation is supported by ample historical evidence. But then the politicians go on to conclude that research budgets in the universities should be cut and that they should develop a 'customer-relationship' with industry, so that work approved of and financially sponsored by industry would increasingly form the bulk of university research.

This curious and unwarranted conclusion is weak on two counts. First, the fact that exploitation of invention has not materialized in many instances is not the fault of universities (which are not equipped with the commercial and entrepreneurial skills required to launch new products) and the alleged weakness in this respect will not be remedied if the universities cease to generate new ideas and inventions.

Second, if research work at universities could only be carried out on the approval of industrial sponsors, then many novel ideas and long-term research projects would never see the light of day, because the curse of short-termism which is so prevalent in industry is likely to dominate. Furthermore, confining university research to industrial sponsorship would restrict the freedom of research workers from publishing the results of their work and could become a serious impediment to the dissemination of knowledge for the benefit of industry as a whole.

Delegating to industry the responsibility for financing and supervising university research is not the way to cure the problem of failing to exploit inventions. It is akin to curing a disease by killing the patient.

MNEs and sovereign states

The question of the relationship between MNEs and host countries is discussed in the previous three chapters. The level of mutual suspicion and hostility, while understandable historically, is inhibiting progress of international trade and many opportunities for new ventures and economic improvements in the host countries are lost.

In seeking to encourage governments to adopt a policy that is nearer to a middle ground, where reason might replace rhetoric, it is appropriate to mention an interesting symposium in Washington DC to mark the fiftieth anniversary of the US Export-Import Bank, where a collection of papers by distinguished academic economists was assembled under the promising title 'Strategic Trade Policy and the New International Economics'. In the introductory paper entitled 'New Thinking About Trade Policy', Krugman (1986) states:

> The traditional faith in the efficacy of the market partly reflected a judgement about reality; equally it reflected a lack of any ability to describe precisely

what difference deviations from perfect market makes. On one side, we are forced to recognize that the industries that account for much of world trade are not at all well described by the supply and demand analysis that lies behind the assertion that markets are best left to themselves. As we have seen, much of the trade appears to require an explanation in terms of economies of scale, learning curves and the dynamics of innovation – all phenomena incompatible with the kind of idealizations under which free trade is always the best policy. Economists refer to such phenomena as 'market imperfections', a term that in itself conveys the presumption that these are marginal to a system that approaches ideal performance fairly closely. *In reality, however, it may be that imperfections are the rule rather than the exception.*

The imperfections, as he calls them, that we have identified in this book could, taken to an extreme, be formidable. Consider the situation in which the top of the hierarchy of an industrial sector (power generation, for instance) is dominated by a single figure of supranational enterprises, themselves further linked by alliances in particular elements of their products and markets. The location of their headquarters – their nationality so to speak – may no longer be where their principal manufacturing activities are located but may be selected for geographical, fiscal or even climatic attraction. Sovereign states needing to have a significant element in their national economies (accepting that nations need a 'balanced portfolio' for economic stability) would then need to compete, in order to attract these macroindustrial enterprises to establish subsidiaries in their country.

The particular skills, costs and attitudes of labour, the availability of support industries, subsidized start-up, and the all-important stability of governments and their policy in respect of industry – all these would be important factors in such competition. This, if it happens, would bear no resemblance to the 'wealth of nations' model of the past, with its busy national microenterprises engaged in price competition to supply similar products to a national market. The credibility of the conventional international trade model, based on the comparative advantage of nations in specific product areas, must also come under question.

This admittedly Orwellian scenario is not alone in the challenge it presents to the concept of sovereignty of nations. Nation-centred models of many aspects of human affairs are all losing their validity. Pollution, the problems of the Third World, the consequences of misuse of nuclear technology and advances in biology and biochemistry are forcing the recognition and acceptance that we live in a global village and that national interests must become subservient to some global consensus. The world of the twenty-first century will be, in communications terms, as closely knit and interdependent as were the towns of a single nation at the end of the nineteenth century.

To be set against this are distinct national cultures and the responsibilities of governments as the guardians of the interests of their electors. Since industry constitutes a vital element in the economy of a country, even the most committed free market politician feels obligated to act. If the tried measures of state ownership and subsidy of selected industrial sectors are to be discredited, what else can and should be done?

A code of conduct

The base line from which to build such a policy would seem to be the degree of state intervention that has been tried and taken for granted. Through regulation, and through custom and practice, industry has accepted that its financial conduct, its products, its treatment of employees and its impact on the life of the society in which it operates should meet minimum standards. We suggest that the way forward is to seek international agreements to reproduce and enforce such minimum standards to be observed by the subsidiaries of all MNEs. The objective would be to achieve the same involvement and commitment to the host country as is exhibited by its nationals. In exchange, such subsidiaries would be afforded identical rights and treatment to those afforded to national enterprises – full 'citizen' status, as it were, not just a 'work permit'.

Although efforts have been made by UN and other agencies to tackle the problems of relationships between MNEs and sovereign states, many remain unresolved. It is clear that these efforts should continue, and in particular, we believe, in the following three directions, as discussed in Chapter 10:

- There is a need for a charter by host countries to allay the fears of MNEs, particularly in relation to nationalization and expropriation of assets without full and speedy compensation through a binding independent arbitration mechanism, but also to protect MNEs from arbitrary changes in the rules of the game.
- There is a need for MNEs to subscribe to a code of conduct that will protect the host country and its inhabitants from exploitation, malpractices, damage to the environment and loss of sovereignty.
- MNEs should strive to operate through joint ventures with local operators and thereby eliminate some of the conflicts between local interests and those of the MNEs.

The purpose of the first two measures, and the difficulties in implementing them, are self-evident. There is no doubt that efforts to accelerate this implementation would be in the long-term interests of all concerned.

But progress would depend on creating a critical mass of countries and MNEs prepared to proceed in this way. This would then lead to strengthening the rule of international law and would put pressure on other countries and MNEs to join. Inevitably, in some countries progress will be impeded by political instability and serious internal strife.

The third measure listed above is not one that many MNEs would favour, because of the potential loss of freedom of action and dilution of managerial control, discussed in Chapter 10. And yet, in many countries this may well be the most effective way of eliminating conflict and potential hostility. The more success stories of joint ventures come to light, the more convincing and attractive would such a solution become to both sides.

As the discussion in Chapters 9 and 10 clearly shows, the difficulties and real conflicts that can arise are formidable. The conditions required for a foreign subsidiary 'to be like an indigenous company' are:

1 It should be a true industrial entity with all the functions of design, development, production and marketing present.
2 The job opportunities it offers should be open to nationals of the host country at all levels with the possibility of appointment to the top of its global corporate structure.
3 Consequently, the corporate headquarters should be multinational and avoid the appearance of bias towards nationals of the parent company.
4 It should, through a mixture of formal reports and publicity, reproduce the counterpart of the statutory reports and accounts, to enable the same public scrutiny as for publicly quoted national companies. Depending on the degree of local autonomy on public matters, these statements should be backed by the authority – and physical presence – of top officials of the enterprise.

Would these provisions impose constraints that are incompatible with the value-creating capabilities on which the case for big MNEs rests? Is it too much to ask of companies operating internationally, some with only a limited degree of local manufacture in a few host countries?

First, it is necessary to remind ourselves that our concern is limited to the macroindustrial scene, involving some hundred or so companies worldwide with annual sales of many billions of pounds. Only these enterprises are capable of becoming a significant force in the economy of host nations and thereby provoke concern. Second, these requirements could be used as the basis for a specific framework within which any outstanding problems could be debated and resolved. Finally, the reality of global enterprises such as Ford, General Motors and IBM is not far off meeting these requirements.

The state, in return, should treat the companies as if they were national enterprises. Currently, whether in civil disputes, tax collection or even local wars, the MNE often finds itself in the middle, caught between opposing views of host and parent states. Particularly the latter, often influenced by pressure groups, finds it tempting to use its influence over corporate headquarters to apply political pressure on the host government. However, the host nation has not been averse to exerting pressure in the opposite direction. The more powerful the state the more likely it is to put extraterritorial pressure on other countries. The US has been one of the most active in using this weapon in global foreign policy. There can be no question but that it is to be deplored in the economy of the future. Generally, the evidence shows that MNEs are opposed to it, preferring the benefits of internationalism to the disadvantages of being branded as agents of their parent governments.

The global issues

This brings us to the major unresolved issue of the application of 'competition policy' to the operations of MNEs. As we have seen in the case of GEC and Plessey, national bodies such as the MMC (Monopolies and Mergers Commission) in the UK are in some difficulty when faced with transnational issues. It could even be said that countries with a vigorous anti-trust policy are so diligent in preventing the achievement of a 'critical mass' by mergers of their national company that they become instead takeover fodder for foreign companies not so restricted.

The President of the German Federal Control Office went further in opening an anti-trust conference in Berlin in June 1990, expressing concern that the sort of alliances described in earlier chapters were potentially 'international cartels', which could lead to serious deformation of the world economy. The logical case for a global rather than a national approach to competition in the case of macroindustrial enterprises and their alliances is persuasive. If it is accepted that an overconcentration of market strength on a national scale is to be prevented, it must follow that the much greater strength of global dominance is an even bigger danger to society and to consumers.

Thus, the challenge to states comes from several directions but can only be met by international accords, in which the future development of MNEs is encouraged, while the dangers to the public from abuse of their massive economic and market power are strictly controlled. Success in this direction would not only bring benefits to the industrialized world in which they operate, but have a potential to be exploited in attacking the problems of the Third World.

An important aspect of these problems is the dependence of Third World countries on the uncertain income from such assets as nature has given them – agriculture, minerals, climate, etc. – and their lack of 'appropriate' industries in their economies. Attempts to redress this imbalance by lending them money for development are rarely accompanied by help on how to spend it effectively. Peter Drucker (1981), in an essay entitled 'Multinationals and Developing Countries – Myths and Realities', writes:

> The best hope for developing countries, both to attain political and cultural manhood and to obtain the employment opportunities and export earnings they need, is through the integrative power of the world economy. And their tool, if they are only willing to use it, is, above all, the multinational company – precisely because it represents a global economy and cuts across national boundaries.

The unwillingness he refers to is, in large measure, fear of exploitation and dominance by an enterprise with financial resources that exceed many times those of the developing country. This fear can only be allayed if MNEs behave responsibly and if their actions are seen to be regulated by an international body to conform to international standards (such as we propose), and with the assurance that there is redress if they are not complied with. In return, the MNEs must be assured by this international body, of similar redress and compensation if, through civil unrest or political action, its assets are confiscated or operations suspended. If such arrangements can be achieved, a powerful tool would be created to replace the current adversarial climate, described by some as leading to 'rich nations getting richer, while the poor are getting poorer'.

Thus, the issues facing industry and national governments are big, and inevitably they are global. They represent a serious challenge to the traditional concepts of nationality and sovereignty, from which the model of the world economy is derived. Progress on these issues can only be made by a mixture of pragmatic opportunism and retention of long-term goals. The general problem of ordering our global economy has been on the agenda for most of the twentieth century. How is real progress to be made?

First, it seems that nations with the greatest economic strength at any particular time are unlikely to take the initiative. Their seemingly unassailable success, and consequently their satisfaction with the status quo, makes the apparent surrender of their strong position through regulatory agreements unlikely in the face of national political criticism.

The USA was in this position until recently and, in its political pronouncements, there remains a desire in influential industrial sectors to re-establish this dominance. Now that Japan has taken over, we find that its industrialists exhibit in their public pronouncements a global, if Japan-oriented, vision. Unfortunately, Japan's unique cultural background puts it in an unpromising position to understand, and be understood by, other nations.

The most promising basis on which to build the required new approach to supranational understanding in general is the EC. Fears until recently that its goal was a self-satisfied, inward-looking European nationalism, isolating itself from the rest of the world, have been greatly reduced by the appearance of the east European market and by the dramatic unification of Germany. Policies embracing these problems within the original '1992 concept' of the EC have to embrace many of the global issues discussed in this book. Furthermore, if a combined USA and EC approach is made to tackle the industrial problems of the member states of the Soviet Union and its previous satellites, an ever wider coverage of global issues will have been addressed to become a powerful instrument in addressing the remaining global problems.

These are challenges that concern governments and industry. They should be faced without further delay.

References

Drucker, P. F. (1981), 'Multinationals and Developing Countries – Myths and Realities,' in *Towards the Next Economics and Other Essays*, Butterworth-Heinemann.

Frosch, R. A. (1983), 'On Tap but Not on Top,' in *Man and Technology*, Cambridge Information and Research Services.

Krugman, P. R. (1986), 'New Thinking About Trade Policy,' in *Strategic Trade Policy and the New International Economics*, MIT Press.

Index

ABB, 48, 53, 54
ACEC, 54
Acorn Computers, 25
Acquisitions, see Mergers
Admiralty Research Laboratory, 3
AEG, 54
Aero-engines, 46, 200
Aero-industry, see Aircraft industry
Aérospatiale, 36, 39, 127
Aggregation, see Mergers
Agroindustrial technology, 139
Agusta, 36, 37
AIM, 134
Airbus, 32
Airbus Industrie, 126
Aircraft industry, 3, 7, 20, 30–42, 45, 46, 82, 114, 199
Airliner, see Aircraft industry
Alcatel, 57
Allende, Salvador, 182
Alliances, 49, 50, 51, 59, 104–8, 141, 201
Allied Signal, 48
Alsthom, 53, 55
Alvey Directorate, 130
Alvey programme, 117–19
American Home Products, 56
American industry, see USA
Annual reports, 18, 29, 42, 176
Ansaldo, 54
Anti-trust, see Monopoly
Apollo programme, 128
Applied research, 5, 9, 137
'Arandee', see R & D
ARC, 103
Argentina, 191
Arianespace, 128
Arm's length, 151, 193
Armament Research Department, 3
ASAB, 53
Asea, 53
Aston Martin, 163
AT&T, 48
Audi, 164
Audretsch DB, 138, 144
Australia, 120, 165
Autolatina, 51
Automobile, components, 51
Automobiles, 20, 45, 46, 49–51, 107–8, 136, 161, 162–6, 189, 199

Baillieu CC, 57
Balance of payments, 170, 183
Balance sheet, 152, 176
Banks, 102, 103
Bardach, E., 197
Barnevik, Percy, 54
BASF, 48
Basic research, 5, 9, 129
BAT, 47, 152
Baumol, W. J., 86, 89
Bayer, 48, 56
BBC, 23, 42, 110
Beaird-Poulon, 52
Beechams, 23, 56, see also Smith Kline Beecham
Beer, 68
Bell Helicopter, 39
Bendix, 53
Bergsten, C. F., 197
Bergsten et al, 196
Besse, 125
Beta-max, 200
'Betting the company', 23–42, 33, 35, 199, 201
Betts, G. G., 81, 89
Bhopal, 66
Bid-proof, 150
Bide, Sir Austin, 118
'Big is necessary', viii, 62–90, see also Size
Bilateral relations, 155, 178, 181–2
Biochemistry, 139
Biological resources, 139
Black, Joseph, 2
Bleakley, M., 49, 50, 59, 60, 107–8
BMW, 48
Boeing, 31–3, 48, 82, 121, 127
Bootstrap option, 100–1, 105
Bosch, 48, 51
Boveri, Walter, 54
Brain drain, 137
Braudel, F., 1, 21
Brazil, 159
Break even, 19, 27, 77
Bridgestone, 51–2
Brigham, E. F., 70, 74–5, 90
Bristol Aeroplane Co, 34, 36, 126
Bristol Myers, 56
Bristol-Siddley, 34
Bristow, Alan, 41
Britain, see UK
BRITE, 138
British Aerospace, 48, 119, 127, 151
British Aircraft Corp, 31–2
British Airways, 37
British Association, 2
British Petroleum, 47, 67, 152
British Rail Engineering Ltd (Brel), 54
British Telecom, 47, 117

Index

Brown Boveri, 53
Brown, Charles, 53
BTR, 48, 188, 189
'Build-own-transfer', 121
Bush, George (President), 189
Business case, 11–13
Buzacott, J. A. et al, 80, 82, 89

C. Itoh, 47
California, 174
Calvo doctrine, 192
Camcorder, 20, 45, 199
Canada, 120, 164, 165, 188
Cantley, M. E., 89
Capacity, see Size
Capital, 147, 168, 172, 173, 174, 176, 178, 183, 184–5
Capital intensive plant, 85–6
Carbodies, 163
Cartel Office, 149
CASA, 127
Cash flow, 16–21, 27, 28, 39, 201
Caterpillar, 48
Caves, R. E., 170, 177
Cayman Islands, 152
CEC, 132, 137, 138, 140, 141, 142, 143, 144, 151, 201
CEGELEC, 55
Central America, 154
Centralization, 65–6
Centres of destination, 168–9
Centres of manufacture, 167–9
Centres of origin, 167–9
CERN, 138
CFIUS (Committee on Foreign Investment), 150
CGCT, 58
CGE, see Compagnie Générale d'Electricité
Channel Tunnel, see Eurotunnel
Characteristic price, see Price
Characteristic volume, see Volume
Chemicals, 161
Chile, 182, 191
Chirac, Jacques, 125
Chrysler, 48, 164
CIA 182
Ciba-Geigy, 48, 53, 56
Citroën, 164
Civil aircraft, see also Aircraft industry
Coca-Cola, 187, 188
Code of conduct, 195–6, 208
Coleman, R., 116, 130
Columbia, 191
Combustion Engineering, 54
Commission of the European Communities, see CEC (see also EC)
Committee of Review on Offsets, 130
Commodity market, 21
Compagnie Générale d'Electricité 48, 54, 55, 57, 124

Company size, see Size
Competition, 91–108, 112–13, 115, 142, 143, 145, 148–51, 172, 173, 174–6, 187, 198, 210, see also Monopoly
Competition Council, 150
Competitive advantage, 92, 100, 142
Computers, 20, 25, 44, 45, 46, 98, 134, 199
Congress, 150
Consensus, 126, 129
Consolidated Goldfields, 100
Continental/General Tire, 52
Control Data Corp, 123
Convair, 127
Cooper, 52
Copernican revolution, 155
Coplin, J. F., 12, 21
Copyright, see Intellectual property
Corfield, Sir Kenneth, 26
Corporate challenge, 91–108
Corporate networks, 190
Corporate performance, 63
Cost, 172, 176, see also Unit cost
Cost-plus, 114
Credibility, 28–30
Credit, 28–30, 182
Crédit mixte, see Mixed credit
Critical mass, 35, 42–61, 88, 199–200
Cuba, 191

d'Estaing, Giscard, 125
D-ram microchips, 154
Daimler-Benz, 48, 127, 150
Damon, 123
Dathe, J. M., 66, 81, 89
de Haviland, 31, 126, 127
De Lorean, 163
Dean, Joel, 70, 82–5, 89
Defence procurement, see Public purchasing
'Dehumanized workforce', 63
DeLamarter, R. T., 159, 177
Delta, 134
Demand, see Volume
Denmark, 165
Department of Commerce (US), 159, 177
Department of Justice (US), 149
Desgorges, Jean-Pierre, 55
Destination country, 178–81
Development grants, 191
Devlin, G., 49, 50, 59, 60, 107–8
Diesel locomotives, 99
Diffusion of technology, 140
Digital (DEC), 48, 123
Digital switch, see Telecommunications
Dissemination of knowledge, 139
Distribution, 99, 101, 178
Dixons, 25
DIYE (do-it-yourself economics), 110
Domestic appliances, 52–3

Douglas, 32, 127
Drive, 134
Drucker, Peter, 211, 212
DTI (Department of Trade and Industry, UK), 116, 117
Du Pont, 48, 56
Dukakis, Michael (Governor), 189
Dumping, 153, 154
Dunlop, 51–2
Dunning, J. H., 160, 177

E H Industries, 37
Early, Joseph, 189
East India Company, 159
Eastman Kodak, 48
EC, viii, x, 58–9, 61, 92, 115, 122, 131–44, 151, 155, 163, 164, 171, 212
EC policy on innovation, 131–44
ECIL, 123
École des Mines, 125
École Normale d'Administration, 125
École Polytechnique, 125, 156
Economic blocs, 131
Economies of scale, 43–60, 68–89, 161, 201
Edelstein, M., 159, 177
Education, 110–13, 125, 129
Educational system, *see* Education
Edwards, R. S., 80–1, 89
EEC, *see* EC
Egypt, 191
Eilon, S., 65, 89, 102, 108
Electric power, 53–5
Electrolux, 48, 52–3
Electronics, 138, 189
Elf Aquitaine, 47
Eli Lilley, 56
Emerson Electric, 48, 52
Employment, 172, 173
Enabling technologies, 141
Energy, 119, 139, 140
Engineers, *see* Scientists and engineers
ENI, 160
Environment, 66–7, 139, 140, 141, 155, 175, 178
Ericsson, 57, 58
ESF, 138
ESO, 138
ESPRIT, 118, 134, 138
Ethylene plants, 84
EURAM, 138
European Community, *see* EC
European Gas Turbine Co, 55
European Parliament, 140
European Space Agency (ESA), 128, 138
Eurotunnel, 55, 129, 146
Exchange controls, 193
Exchange rates, 93
Exon-Florio, 150
Extraction, 178

Exxon, 47, 160

Fabius, Laurent, 125
Failures, 20, 23, 26, 30–42, 65–6, 199
Fairey, 36
Feasibility, 9, 10, 15, 198
Fellowship of Engineering, 107
Finmeccanica, 54
Finniston Committee, 111, 130
Firestone, 51–2
Fisheries, 139
Fission, 139
Fixed costs, 68, 70–1, 85–6
Fleming, Sir Alexander, 55
Flymo, 52
Ford, 48, 49, 51, 151, 160, 162, 163, 164, 188, 189, 190, 210
Fortune, 162
Framework programme, 138–41
France, vii, 29, 30, 36, 37, 51, 121, 170
Free market economy, 109–10, 144, 153, 203
French industries, *see* France
Frescati rules, 112
Friedman, A., 98, 108
Frosch, R. A., 205, 212
Fujitsu, 48, 146

Gamble, *see* Risk
Gandhi, Indira, 38
Gas storage holders, 80–1
GATT, 153, 182
GEC, 24, 26, 48, 53, 54, 55, 57, 58, 60, 117, 149
General Dynamics, 48, 121
General Electric, 48, 52, 54, 55
General Motors, 48, 49, 51, 52, 67, 160, 164, 205, 210
General Tire, 52
German industry, *see* Germany
Germany, vii, 29, 30, 36, 51, 92, 111, 121, 149, 160, 189, 204, 212
GKN, 149
Gladwin, T. N., 146, 156
Glagolev, V., 89
Glaxo, 55, 56
Global competition, *see* Competition
Globalization, 145, 146, 155–6
GNP, 140
Goals, 183–9
Gold, Bela, 68, 78, 89
Golden share, 124, 151, 188
Goldring, Mary, 23

Index

Goldsmith, Sir James, 30
Goodrich, 52
Goodyear, 48, 52
Government research, 2
Government support, viii, 30–42, 201, 203
Governments, 109–30
Governments and innovation, 109–30
GPT, 58
GrandesÉcoles, 111
Great Depression, 109
Grindley, P., 130
Grosse, R. E., 196, 197
Groupement d'Intérêt Economique (GIE), 127
Growth of MNEs, 160–2

Hanson, Lord, 188
Hanson Trust, 30, 100
Harmonization, 117
Harvard Business School, 145
Harvey-Jones, Sir John, 146, 156
Hawker-Siddley, 32, 34
Hawthorne, E. P., 98, 99, 108
Health services, 56, 139
Hedging, 27, 28
Helicopters, 36–42
Henderson, David, 110, 130
Henderson, Sir Denys, 24, 26
Henri IV lycée, 125
Herfundahl-Harschmann Concentration Index, 148
Hertner, P., 159, 170, 177
Hewlett-Packard, 48
Hilger Instruments, 25
Hitachi, 48, 54, 122, 146
Hitch, C. J., 4, 21
Hoechst, 48, 56
Home country, 173–4, 178–81, 183, 184
Honda, 48, 164, 165
Honeywell, 48
Hong Kong, 173
Horst, T., 197
Host country, 169–71, 174–6, 178–81, 184–9, 202
House of Commons, 130
House of Lords Committee on Overseas Trade, 204
House of Lords Judicial Committee, 152
Hudson Bay Company, 159
Husqvarna, 52

IBM, 48, 65, 67, 123, 187, 188, 210
ICI, 24, 26, 48, 117, 146
ICSID (International Center for Settlement of Investment Disputes), 192
Independent European Programme Group (IEPG), 114
India, 37, 40, 66, 123, 152, 154, 187
Indian Oil and National Gas, 40

Industrial base technology, 173
Industrial property rights, see Intellectual property
Industrial research, 116, see also R&D
Information technology, 117, 132–3, 138, 139, 141, see also Telecommunications
Infrastructure, 173
Ingledon, John, 25
Innovation, 1, 43, 46, 55, 88, 92, 109–44, 156, 168–9, 184–5, 199–200, see also R&D
Innovation, process of, viii, 132
Innovative risk, see Risk
INSID, 196
Intellectual property, 154, 182, 187
Intellectual resources, 141, 143, see also Scientists and engineers
Interest rates, 94
International economy, 155
International law, 192
International trade, see Trade
International trade game, 178–97, see also Trade
Intervention, viii, 95, 109–10, 120, 142–4, 203, 204, 205
Invention, 9, 10, 15, 40, 198
Ireland, 153, 165, 191
IRI, 47
Isuzu, 48, 51
IT '86 Committee, 118
Italy, 36, 37, 39, 170, 191
ITT, 26, 48, 57, 64, 65, 182
IVECO, 51

Jaguar, 51, 151, 163, 188
Japan, vii, 29, 30, 47–50, 53, 54, 59, 111, 121, 134, 135, 140, 160, 163, 164, 165, 166, 170, 171, 204, 212
Japan bashing, 155
Japan Technology Transfer Association, 130
Japanese industry, see Japan
Johnson, Kelly, 11
Joint ventures, viii, 43–61, 107, 143, 194–5, 201
Jones, Aubrey, 113, 130
Jones, G., 159, 170, 177
Junk bonds, 100

Kaske, Karlheinz, 58, 60, 61
Kawasaki, 36
Keith, Lord, 23, 24
Kennedy, Edward (Senator), 188
Kidnapping, 191
Kindersley, Lord, 33, 34
Koontz, James, 126
Koratsu, Hajune, 119
Korea, 121, 159, 173
Krugman, P. R., 207, 212

Labour, 92, 168, 173, 176
Labour intensive plant, 85–6
Lau, L. J., 84, 89
Launch, 7, 9, 13, 15, 17, 18, 27, 35, 43, 44, 50, 60, 70, 96–7
Launch aid, 30, 31, 39, 116, see also Launch
Launch costs, see Launch
Layton, Christopher, 155, 156, 157
LDC (less developed countries), 161, 166, 170, 173, 174, 175, 182, see also NIC; Third World
League of Nations, 109, 130
Learning curve, 13, 207
Lebanon, 191
Level playing fields, 92
Levy, Raymond, 50
Libya, 191
Liechtenstein, 152
Life sciences, 140, 141
Linear cost function, 76–82, 85–6
Linear model for innovation, 3–6
Local content, 153
Lockheed, 10, 35, 48, 127
Lombard, Adrian, 34
London Business School, 118
Long, Brian, 25
Lord, Alan, 20
Lotus, 163
LSI, 98
Lucas, 51

McDonnell Douglas, 48
McKean, R. N., 4, 21
Machine tools, 46
Macmillan Publishing Co, 100
Macroindustrial economy, 200
Macroindustry, 4, 44–6, 48, 149, 200
Macroinnovation, vii, 20, 95–7, 108, 129, 141, 144, 146, 200–3
Madden, C. H., 158, 160, 169, 170, 176, 177
Magglund, 53
MAN, 48, 55, 150
Managing creative disorder, 1–22
Manley, Brian, 24
Mannesman, 48
Manufacturing, 5, 9, 12, 68, 86, 88, 93, 101, 103, 139, 163, 169, 178, 198
Marconi, 2
Marine resources, 139
Market country, 178–81, 184
Market model, 6–9
Market pull, 112
Market research, 115
Market share, 68, 105
Market, marketing, 7–9, 12–15, 88, 91–108, 133, 139, 142, 162–8, 172, 178, 198
Materials, 136, 141
Matra, 58

Matsushita, 48
Maxwell Communications, 100
Mazda, 48, 51
MCW, 163
Mercedes-Benz, 51
Merck, 56
Mergers, 28–9, 43–61, 99–100, 103–4, 105, 148–51, 188
Messerschmitt-Bölkow-Blohm (MBB), 36, 150
Mexico, 191
Michelin, 48, 51–2
Microindustry, 44–6
Microinnovation 95–7, 141
Minerals, 161, 171, 188, 190
Minister for Information Technology (UK), 117, 118
Minister of Science (UK), 4, 21
Ministry of Aviation, 33, 34
MIP, 122
MITI, 119
Mitsubishi, 47, 48, 54
Mitsui, 47
Mixed credit, 123
MMC (Monopolies and Mergers Commission), 57, 58, 61, 149
MNE, x, 147, 158–212
MNE's response, 186
Mobil, 160
MOD, 130
Modernization, 139, 143
Monaco, 152
Monopoly, viii, 48, 54, 56, 64, 149, 184–5, 190, see also Mergers; MMC
Montedison, 48
Moran, T. H., 197
Morrow, Sir Ian, 24
Motor car, see Automobiles
Motorola, 48
Moulton, John, 25
Multinational enterprise, see MNE
Multinational scene, 145–57

NASA, 128
National Audit Office (UK), 118, 130
National champions, 36, 59, 127
National interest, 184–9
National Physical Laboratory, 3
National Westminster Bank, 104
Nationality, 152
Nationalization, 186, 191
Nationalized industries, 126
NATO, 114
NCR, 123
NDC, 173
NEC, 48
NEDO (National Economic Development Office), 55, 130
Negotiations, 181

NEI, 54
Nestlé, 53, 150
Netherlands Antilles, 152
New Deal, 109
New York Times, 188, 197
New Zealand, 165
NIC (newly industrialized countries), 154
Nigeria, 152
Nippondenso, 48, 51
Nissan, 48, 146, 163, 164
Nobel prize, 135
Noblesse d'État, 125
Norton, 188, 189
Norway, 165
Nuclear energy, 139

Oakley, Brian, 117
Observer, 189, 197
OECD, 110, 112, 124, 136, 137, 196, 197
Offset, 120–2
OFT (Office of Fair Trading), 149
Oil, 160, 161, 190
Okimoto, D. T., 119, 130
Olivetti, 25
OPEC, 190
Opel, 48
Origin country, 178–81, 184
Otto, N. A., 2
Ozal, Tugat, 121

Pache, Bernard, 125
Pan Am, 37
Panel on invention and innovation, 21
Panther, 163
Pappas, J. L., 70, 74–5, 90
Parametric costing, 11
Pareto curve, 67
Parkinson-Cowan, 53
Parliamentary Committee on Science and Technology, 118
Patents, 56, 154, 182
Patient money, 20, 33, 42
Peace Shield, 121
Pechiney, 125
Peugeot, 48, 51, 164
Peugeot/Talbot, 162, 163
Pfizer, 56
Pharmaceutical industry, *see* Pharmaceuticals
Pharmaceuticals, 3, 15, 23, 24, 55–6, 154, 161
Philips, 24, 48, 53
Pirates, 154
Planning and control, 168, 171, 172
Platform, 114
Plessey, 57–8, 117, 149
Poitiers affair, 122
Pollution, 144, 207
Polytechnics, 110

Porter, M., 145, 157
Positive role of MNEs, 171
Poverty, 154, *see also* Third World
Power stations, 82
Pratt & Whitney, 34
Price, 7, 19, 44, 46, 188, 194
Price index, 193
Privatization, 54, 124, 143, 203
Product credibility, 7
Product definition, 5, 9, 11–12, 15
Product design, 102, *see also* R&D
Product improvement, 7, 13–14, 15
Product launch, *see* Launch
Product life cycle, 95–7
Product range, 27–8
Production, *see* Manufacturing
Production launch, *see* Launch
Productivity, 68, 69, 74, 92–3
Profit, 40, 152, 176, 186–7, 193
Protectionism, 184–5, *see also* Tariffs
Prototype, 5, 9, 95, 105
Public purchasing, 113–15, 143

Quality, 172
Quality of life, 139
Quotas, 170

R&D, viii, 1, 5, 9, 8–14, 15, 16, 19, 20, 29, 50, 70, 95, 101, 105, 110, 116, 118, 131–44, 161, 171, 172, 178, 194, 195
R&TD, *see* R&D
Racal, 119
RACE, 134
Radar, 3
Railway transport, 54
Rank Group, 25
Raw materials, 91
Raytheon, 48
Regional development, 120, 122, 132
Reith Lectures, 110
Relative costs, 14–16
Reliant, 163
Renault, 48–51, 107, 124, 125, 164
Research intensity, 138
Residency, 152
Restructuring, 131
Rhône-Poulenc, 48, 124, 125
Rich, B. R., 11, 22
Risk, 20, 23–7, 112, 120, 138, 139, 171, 194, 195, 199, 201
Road Research Laboratory, 3
Roberts, Derek, 24
Robock, S. H., 191, 192, 197
Rocard, Michel, 125
Rockwell, 48
Rolls-Royce, 12, 22, 24, 33–6, 42, 54, 151, 163, 188
Roosevelt, Franklin, 109
Roper, 53

Rosegger, G., 81, 90
Rover, 162, 163
Rover-Honda, 51
Rowntree, 150
Royal Aeronautical Society, 6, 11, 21, 32, 42
Royal Aircraft Establishment, 3
Royal Dutch/Shell, 160
Royal Institute of International Affairs, 109
RTD&D, see R&D
RTZ, 125
Russian roulette, 27

S&T, see Science and technology
Saab, 48, 51
St Gobain, 48, 188
Sales volume, see Volume
Sampson, A., 64, 90
Sandoz, 55–6
Sanyo, 48
Saudi Arabia, 121
Saunders-Roe, 36
Scale, see Size
Schap, Anders, 53
Schroders, 25
Schumacher, E. F., 62, 63, 66, 90, 142
Science and technology, 5, 132, 139
Scientists and engineers, 134–5, 137, 140, 205
Scottish Development Agency, 122
Screwdriver plants, 154
SDI, 136
Sears Roebuck, 47
Second World War, 3, 159
Secretary for Trade and Industry (UK), 33, 37, 118
Secretary of Commerce (US), 14
Self-regulation, 174
Shareholders, 29, 49, 150, 174, 188
Sharp, 48
Shell, 47, 67, 160
Short term, 107
Siemens, 48, 55, 58, 60, 149
Sikorski, 39
Silicon Valley, 7
Simmonds, K., 191, 192, 197
Singapore, 173
Single European Act, 131
Single European market, 58
Single market, 131
Size, 20, 43–61, 62–91, 99–108, 199, 202, see also Economies of scale
Skandiaviska Elverk, 53
'Skunk Works', 11
Slide rule, 78
Small companies, 142–3
'Small is beautiful', viii, 62–90, 202, see also Size
Smith Kline Beecham, 56

SNECMA, 34
Social change, 142
Social Charter, 92
Society of British Aircraft Constructors, 34
Society of Motor Manufacturers and Traders, 163, 165, 177
Soft drinks, 68
Software, 46
Soner, 53
Sony, 48, 200
South America, 51, 154, 182, 192
Sovereignty, 147, 155–6, 206
Soviet Union, 212
Space programmes, 128
Space station, 20
Spain, 165, 170
Specialization, 74, 101
Squibb, 56
Standard Oil, 160
Standards, 117, 133, 138, 140, 142
Start-up, 28
State financed companies, 124–6
State intervention, see Intervention
Statistical analysis, 83–4
STC, 26, 119
Steam locomotive, 98–9
Steel plants, 80–1
Steiner, J. E., 32, 33, 42
Stock market crash, 65
Stoleru, Lionel, 155, 157
Stopford, J. M., 160, 177
Strategic alliances, see Alliances
Subsidies, 94, 112, 122–3, 127, 184–5, 191
Subsidized exports, 123–4, see also Subsidies
Sud Aviation, 126
Sulzer, 150
Sumitomo, 47, 51–2
Super 301, 154
Supra-national, 147, 148
Suzuki Motors, 48
Swan, J. W., 2
Sweden, 92
Switzerland, 152
Syria, 191
System X, 57
Systems, 46, 58

Takeover, see Mergers
Tamura, S., 84, 89
Tariffs, 153, 170, 181, 186
Tarmac, 103
Tax, 94, 151–3, 172, 173, 174, 175, 178, 193, 203
Technological 'big bang', 1–3
Technological change, 132
Technological push, 112
Technology, 68, 79, 98, 105, 106, 148, 188
Technology transfer, 120, 178, 186

Tekeda, 56
Telecommunications, 26, 57–60, 107, 132–3, 138, 139, 161, 192
Terrorism, 191
Texaco, 160
Texas Instruments, 48
Thatcher, Margaret, 109
Third World, 123, 139, 208, 211
Thomson, 48, 124
Thorn-EMI, 52
Threats, 91–108
Times, 47, 48, 61
Tomlinson, R. C., 89
Toshiba, 48, 54
Total cost, 71–89
Townsend, H., 80–1, 89
Toyo, 52
Toyota, 48, 51, 146, 164, 165
Trade, 93, 129, 133, 138, 153, 159, 166–71, 202
Trade and Development Enhancement Act, 123
Trade unions, 92, 170, 189–90, 192
Transfer prices, 186, 192–4
Transnational programmes, 138
Transport, 139, 144, 192
Tricity, 53
TRW, 48, 51
Turkey, 121
TVR, 163
Tyres, 51–2

U-shaped cost curve, 73–6, 86–9
UK, 29–31, 51, 54, 57, 111, 114, 119, 121, 146, 160, 162–3
Ulster, 153
UN, 158, 160, 182, 196
UN charter of economic rights, 192
Unbundling, 45
UNCTAD, 182
Unilever, 65, 67
Union Carbide, 125
Uniroyal, 52
Unisys, 48, 123
Unit cost, 71–89
Unitary tax, 151, 174
United Airlines, 65
United Technologies, 41, 48
Universities, 110, 113, 137, 206
Urban deprivation, 175
US Agency for International Development, 123
US Justice Department, 54
US Trade Act, 155
USA, vii, 30, 48, 51, 59, 134, 135, 137, 140, 159, 164, 165, 174, 176, 188, 189, 204, 212
USSR, 128

Valeo, 51
Value-added chain, 168, 178
Variable cost, 71–2, 85–6
Vauxhall, 163
Venezuela, 191
Venture capitalist, 25
Vernon, 176
Vertical integration, 171
VHS, 200
Vickers, 126
Views on MNEs, 166
Volkswagen, 48, 51, 124, 164
Volume, 7, 8, 19, 44, 46, 47, 71–89
Volvo, 48, 49, 50, 51, 107

Wage rates, *see* Labour
Wake Up America, 126
Walter, I., 146, 156
Wang, 123
Water authorities, 54
Watt, James, 2
Wealth creation, 111, 129
Wealth distribution, 129
Weinstock, Lord, 58, 60, 61
Welsh Development Agency, 122
Westinghouse, 48, 54
Westland, 36–42
Whirlpool, 52–3
White Consolidated, 52
White goods, 52
Wilkinson, Brian, 25
Wolfe, P., 86, 89
Work-in-progress, 18
Workforce, 140, 184–5
World Directory of Multinational Enterprises, 160
Wright brothers, 2, 11, 128

Yamani, Sheikh, 121, 130
Yeager, Chuck, 10
Yokohama, 52

Zanussi, 52
Zero-sum game, 107, 168, 191